T0227649

Domestic Energy and Affordable Warmth

Watt Committee Report Number 30

Domestic Energy and Affordable Warmth

Edited by Thomas A. Markus

Emeritus Professor of Building Studies, University of Strathclyde, UK

and

Chairman of the Working Group on Domestic Energy and Affordable Warmth appointed by The Watt Committee on Energy

Report Number 30

Supported by British Gas plc

Published on behalf of The Watt Committee on Energy by

Routledge
Taylor & Francis Group

LONDON AND NEW YORK

First published 1994 by Chapman & Hall

2 Park Square, Milton Park, Abingdon, Oxon OX14 4RN
711 Third Avenue, New York, NY 10017, USA

Routledge is an imprint of the Taylor & Francis Group, an informa business

Transferred to Digital Printing 2004

First issued in hardback 2017

Typeset in 10.5/12 pt Garamond by Julia Stevenson, Brighton

ISBN 978-0-419-20090-1 (pbk)
ISBN 978-1-138-45984-7 (hbk)

A catalogue record for this book is available from the British Library

Contents

Members of the Watt Committee on Energy Working Group on Domestic Energy and Affordable Warmth

This report has been compiled by the Working Group appointed by the Watt Committee on Energy.
The members of the group were as follows.

Authors

Professor Thomas A. Markus	Group Chairman
Dr Brenda Boardman	Environmental Change Unit, University of Oxford
Dr Ian Cooper	Eclipse Research Consultants
William Gillis	Neighbourhood Energy Action
Dr Sonja Hunt	Health Research Consultant
Ann Marno	Energy Action Scotland
Dr Marcus Newborough	Cranfield University
Dr Bill Sheldrick	Heatwise Glasgow

Other contributors

Steve Bendle	South London Family Housing Association
Mike Cooper-Reade	Eastern Electricity plc
Roger Glover	Manweb plc
Carol Godridge	Energy Action Scotland
Richard Lorch	RIBA
Maurice Sharp	British Gas plc
John Stockley	Charted Institute of Building
Colin Sutherland	Chartered Institution of Building Services Engineers

Corresponding members	Leslie Davies	Gas Consumers' Council
	Vincent Hale	British Coal
	Professor Michael Laughton	Queen Mary and Westfield College, University of London
	Peter Sawers	Scottish Power plc
	Norman Worley	British Nuclear Energy Society

Sponsors	British Coal Corporation	British Gas plc
	Eastern Electricity plc	Manweb plc
	Midlands Electricity plc	Scottish Power plc
	as well as	
	Eclipse Research Consultants	Energy Action Scotland
	Environmental Change Unit	Heatwise Glasgow
	Neighbourhood Energy Action	South London Family Housing Association

Editor's note: The working group has thoroughly discussed the questions dealt with in this Report and arrived at a broad consensus for publication. Nevertheless, any view stated here by an individual member of the group is not necessarily the view of the group as a whole nor of any sponsor of the work nor of The Watt Committee on Energy nor of any of its member institutions.

Contributors

Dr Brenda Boardman
Environmental Change Unit
University of Oxford
1A Mansfield Road
Oxford OX1 3TB

Dr Ian Cooper
Partner
Eclipse Research Consultants
121 Arbury Road
Cambridge CB4 2JD

William Gillis
Deputy Director
Neighbourhood Energy Action
St Andrew's House
90–92 Pilgrim Street
Newcastle-upon-Tyne
NE1 6SG

Dr Sonja Hunt
Health Research Consultant
Chadderton Cottage
Bailrigg
Lancaster LA2 0PH

Ann Marno
Director
Energy Action Scotland
21 West Nile Street
Glasgow G1 2GJ

Professor Thomas A. Markus
2 Westbourne Gardens
Glasgow G12 9XD

Dr Marcus Newborough
Department of Applied Energy
Cranfield University
Cranfield
Bedford
Bedfordshire MK43 0AL

Dr William Sheldrick
Heatwise Glasgow
72 Charlotte Street
Glasgow G1 5DW

Foreword

Improvement of the human condition rarely proceeds by sudden or spectacular change, though the total result of slow and irregular changes may be spectacular if today's conditions are compared with those of, say, 30 years ago. To make a real difference to the hardship and waste that arise, a consistent policy, requiring many years to achieve a general improvement, is preferable.

Heating adequate to achieve comfort in the home with minimal trouble on the part of the householder was quite unusual within the memories of many readers of this Report. It has, however, become almost an expectation for the majority of people in the advanced countries. Nevertheless, there remain many British households in which that level of warmth is not achieved – or at least, cannot be achieved within the financial means of the occupiers. As appears at the outset in this Report, the number of the 'fuel poor' in the United Kingdom is apparently some seven or eight million, many of whom are outside the usual definitions of poverty. Already I am using words that must be defined if the size and nature of the problem and the measures that might alleviate it are to be fairly stated.

The sheer size of this situation (representing as it does one-third of British households), the effect of fuel poverty on health and the consequent health care costs, to say nothing of other less tangible but very real hardships, makes it a substantial burden on society – seemingly more than in some other industrialized countries, even if their climates are much colder. These must be the basic justifications for undertaking the study that is reported here. The trigger for this study actually arose, however, from concerns that were expressed at the conference on 'Energy, buildings and the environment', run by the Watt Committee at the University of Strathclyde, Glasgow, in 1991, after studies by a previous Working Group whose Chairman was Professor Victor Torrance.

This topic, and the study whose results are published in this Report, therefore marks a change of emphasis in the interests of the Watt Committee in recent years. Reference to the list of previous Watt Committee Reports (at the end of this volume) shows that when the Watt Committee was formed in 1976 as a response to the oil crises of the early 1970s, it was mainly interested in energy resources. By around 1980, the Watt Committee was turning to environmental questions – the attitude of our Executive at that time was swayed by Kenneth Mellanby, to whom, after his death some months ago, I am glad to pay this belated tribute. I like to think that, through our work on acid rain and related subjects, much of which he led, we played our part in the great change in public attitudes that has occurred in the last 15 years or so, culminating in the Government's commitment, made in Rio de Janeiro, to a long-term programme of environmental improvement, largely dependent on improving the energy efficiency of our housing stock, business and transport.

The other theme of the early Watt Committee Reports was the **rational use of energy** – a term which, at that time and since, we have preferred to 'energy conservation' or 'energy efficiency'. This work on fuel poverty illustrates the reason for the preference. A programme of measures to alleviate fuel poverty may well increase the consumption of fuel in some households, though they would hardly affect the UK commitment to the Rio target, and the increase in atmospheric pollution would be negligible in the world context. Considered as a contribution to human happiness, however, and as a reduction in the strains caused by health problems and defective housing, notwithstanding the possible effect on the depletion of resources and on energy conservation, these measures would surely be **rational**. In the UK, the charging of Value Added Tax on fuel, although it has attracted so much controversy, seems likely to have some effect on the extent of fuel poverty. For the purpose of this study, therefore, environmental improvement (in the usual sense) takes second place to the reduction of fuel poverty.

As I have remarked, the Watt Committee has considered many key questions regarding both security of energy supply and the environmental implications of the world's dependence on energy. Other issues that deserve increasing attention are economic and social aspects of energy production and use. Looking farther ahead, therefore, as one of the projects in our future programme, we intend to review all these issues as the background to energy policy decisions whose effects will perhaps not become evident for half a century. For this study we have adopted the title 'Energy now and the next fifty years', and we would welcome expressions of view and offers of help.

Like these much larger questions, the contents of the present Report benefit from the very wide range of subject skills represented on the Watt Committee. As always, the objective of the Working Group has been to present an authoritative and unbiased view of the facts, and to make recommendations only where the facts – as distinct from preconceived attitudes – justify them. The members of the Working Group, serving as volunteers, have placed their professional knowledge at the Watt Committee's disposal and have found time in their busy lives to take part with enthusiasm. The work could not have been done without financial support by sponsors, whose practical experience has also been valuable to the group. All these are listed in their due places, and to them all, especially to Tom Markus as Chairman of the Working Group, on behalf of the Watt Committee I am glad to express my sincere thanks.

G.K.C. Pardoe
Chairman, The Watt Committee on Energy,
May 1994

Overview and policy recommendations 1

Thomas A. Markus

The Watt Committee on Energy was conceived in the mid 1970s as a channel for discussion of questions concerning energy in the professional institutions. In 1992 it set up a Working Group to examine the set of issues relating to domestic energy use, hard-to-heat houses, low income and fuel poverty. Since starting its work the problems facing low income households have been exacerbated by the imposition of VAT on domestic fuel. This document is the final report of the Working Group.

These issues have been the subject of numerous studies, parliamentary enquiries, academic research and publications in the last two decades. Government departments have been involved, as have suppliers of fuel, local authorities and housing associations; there are major agencies carrying out thermal improvements using government funding; and several voluntary organizations act as pressure and campaigning groups. Therefore it can legitimately be asked whether the Watt Committee can add anything new to this discussion, or formulate new policies to tackle the issue.

The answer is that the Watt Committee, as the result of its constitution, is in a uniquely independent position and thus able to attain a degree of objectivity that is diffficult for these other institutions, though many of them were represented on the Working Group or presented evidence to it. It has been able to review the evidence, establish a clear methodology for considering policy options and propose a range of solutions.

The Group has accepted evidence that around 8 million British households, in receipt of one or more social security benefits, are unable to achieve the comfortable, safe and healthy temperatures which they require in their homes throughout the year. But the problem affects many more people than those classified by such official indices as being 'poor'; the additional number form a second category which can be said to suffer from '**fuel poverty**'. There is a third category of the '**nearly fuel poor**' who are only just able to obtain adequate warmth in normal winters. For them, any increase in fuel costs –

such as that caused by a severe winter, or that caused by the imposition of VAT on domestic fuel without full compensation for low-income households – places them into the category of 'fuel poverty'.

There can be a number of reasons why a household cannot afford adequate warmth:

- Cold climate or exceptionally severe weather.
- Low income.
- Lack of access to an economic fuel.
- A poorly insulated house and/or excessive air leakage.
- Absence of or inadequate heating system.
- Special heating needs, such as longer-than-normal heating periods or the need to meet continuous full comfort temperatures, as in the case of the chronically sick, the elderly, the disabled, young children and unemployed people.

The consequences of cold houses include:

- Discomfort.
- Cold-induced illness, or exacerbation of existing chronic illnesses.
- Excess winter or cold-weather mortality.
- Hypothermia.
- 'Spatial shrink', whereby only one heatable room is used for eating, entertainment, study, children's play and even sleeping.
- Disconnections and fuel indebtedness.
- Condensation on cold surfaces, with consequent mould growth, causing:

  mould-spore induced illnesses;
  emotional and psychological illness;
  social deprivation resulting from a reluctance to invite visitors to the house due to its discomfort, its appearance and its damp smell;
  deterioration of furnishings, interior decorations, clothes, toys and household equipment;
  deterioration of building fabric.

There is evidence that regions with much colder climates than that of the UK – e.g. the Scandinavian countries, Canada, the northern states of the USA and northern Germany – suffer to a much lesser degree from underheated houses with all the attendant consequences.

In an attempt to overcome their inability to heat houses adequately, low-income families in the UK spend a far greater proportion of their total household expenditure on fuel than the average. In 1991, for instance, their percentage fuel expenditure was between 2 and 2.5 times higher than the national average (4.7%), rising to as much as 13%. For low-income pensioners and single-parent families the figure was even higher.

There are two related groups of factors at work: the first concerns the housing stock; the second, social and economic conditions. The energy inefficiencies in the housing stock can be quantified by using a rating such as the **National Home Energy Rating (NHER)** obtained by calculation through an energy audit. The households at risk of fuel poverty, through social and economic conditions, include the unemployed, the elderly, those on low incomes – especially large and single-parent families – and the disabled and chronically sick. As a result of market forces and letting practices there is a strong association in both the public and the private sectors between energy inefficient houses and households at risk, with the consequence that those with the least resources find themselves, in general, in the most difficult-to-heat houses. By establishing the relationship between house condition data and socio-economic data the strength of this association could be measured. Although it is believed that the necessary information for doing this is available within the English House Condition Survey, such analysis has not so far been carried out. Similar unpublished data may be available for Scotland.

Chapter 2 provides the evidence on the nature and magnitude of the problem.

1.3 Costs and benefits

In considering the allocation of resources to tackle underheating the Group has identified two models: cost effectiveness and cost benefit. The first model relates the cost of investment to quantifiable and measurable savings – mainly reductions in fuel expenditure – usually over a fixed term of years. Internal rate of return and payback period are but two commonly used measures of cost effectiveness. The second model evaluates all the costs and all the benefits of a policy, including the costs of not doing something, over the lifetime of the system. One of the problems associated with the cost benefit model is that of determining the scale of the system. For instance, should an attempt be made to evaluate the global environmental effects?

Whichever model is used, there is a basic question of system boundaries and externalities: cost effectiveness or cost benefit to whom? Is it the individual, the household, the local community, the local authority, the government, the fuel industries, or even the entire global community (if, for instance, climate change, pollution or reduction in fossil fuel stocks are considered)?

Many of the consequences of cold and damp houses are easy to see in terms of illness, suffering and loss in quality of life, but difficult to quantify in cash terms. The inefficient use of fuel contributes significantly to CO_2 emissions. The direct costs to the health services of treating condensation-related illnesses has been estimated at £800 million per annum; if cold-related illness is included the figure is nearer to £1000 million per annum. The costs in lost work and schooltime must be equally large but are difficult to estimate. The costs of repair, maintenance, insurance claims, replacement of damaged goods and the loss in capital value of housing property, are substantial and are borne by both the public and private purse.

There is an unspecified amount for fuel in Income Support but it certainly does not take into account the true cost of heating the types of house occupied by most recipients of the benefit. Only if it did so would progressive thermal improvements to such houses justify a reduction in Income Support. In reality what will happen is that such households will choose to take up all or some of the benefits resulting from energy improvements in achieving warmer conditions. Until such time as these take place, additional income will be needed.

Reducing the energy inefficiency of the UK housing stock also has employment implications, since insulation, draughtproofing and the installation or improvement of heating systems will have an impact on both manufacturing industry and the availability of semi-skilled and skilled labour opportunities in the building industry.

Attempts to quantify costs and benefits have foundered for lack of reliable data. There are few surveys of house temperatures. Systematic nationwide auditing has yet to take place, though a few local authorities and other agencies have made a start. There is some evidence from house condition surveys on building quality and heating, but it is not combined nor can it be related to household income. To date it has been impossible to establish the strength of the relationship between the worst quality housing and the lowest income households. It is crucial that the poorest people can purchase the cheapest warmth; therefore their houses require the highest energy efficiency. To achieve this will require a measure of positive discrimination. The lack of data concerning the costs of such action and of the resultant benefits means that estimates have to be used. To remedy this a full audit of fuel poverty is needed. The Group regards such an analysis as an essential component of action – whatever policies are pursued. But even before this is carried out it is evident that the true costs of capital investment and short-term cash support for fuel in low-income households are far smaller, if the benefits are evaluated, than is commonly said to be the case.

Chapter 8 explores methods for evaluating the costs and energy savings of different policies but policy evaluation cannot be restricted to such cost-effective analysis: it must be based on cost benefit.

1.4 Current policies and initiatives

A number of current government housing and energy policies have a bearing on the issue of affordable warmth, but there is no evidence of an integrated interdepartmental policy which addresses this specific issue, nor indeed of an integrated energy policy underpinning this. The most articulated policy is the commitment to reduce CO_2 emissions. This has an indirect effect in that it involves energy efficiency measures and switches of fuel. Some policies pursued by the **Energy Savings Trust (EST)** and the switch to more efficient lighting and domestic equipment will effect a direct reduction in CO_2 emissions (Chapter 4 discusses the energy performance of equipment). However, this policy's short-term impact on fuel consumption for space heating in low-income households, which contribute 24% of the total emissions from the domestic sector, is slight as any increase in energy efficiency here will be used to attain better house conditions rather than to reduce the amount of fuel used. Indeed there is evidence that there can even be a modest increase in consumption; for instance, an 18% increase in gas consumption was measured in DoE's Green House programme. In the longer term, of course, the effect on emissions will be more significant. Switching to cheaper heating fuels or lower tariffs for an existing fuel or increasing fuel allowances in Income Support would increase fuel consumption and CO_2 emissions. These considerations point to the need to develop special policies with respect to low-income households as part of the general strategy to reduce domestic CO_2 emissions.

If warmth is regarded as a benefit, then low-income households are the most effective policy target, as is the choice of tackling the least energy efficient stock.

The government's energy efficiency and housing policies have resulted in various programmes, which the Group has reviewed. Some of these are energy-specific; in others, energy is often only a minor part of general investment in new housing or refurbishment. The programmes include the **Housing Investment Programme (HIP)**, **Home Energy Efficiency Scheme (HEES)**, the **Green House Programme**, **Estate Action**, various house improvement grants and the partnership with the gas and electricity industries in funding the EST. The Group has assessed the degree to which these can meet the demand for investment in thermal upgrading of houses occupied by low-income households and found that not only do they fall far short of meeting existing problems, but they will not even prevent further deterioration (Chapter 3).

In principle, it is widely agreed that energy auditing and rating of the UK's housing stock is an important method of targeting needs to form a part of any strategy in the energy upgrading of the stock. There is some debate about the most suitable rating method. Although the Group has accepted the NHER system as being the most reliable, a change to any other method does not invalidate the argument for rating. Rating would make it possible to allocate capital funds according to identified priorities based on energy efficiencies and to follow this by continuous monitoring of changes in the thermal efficiency of the stock. It would also enable social security and other benefits to be allocated on the basis of the energy performance and, hence, fuel demands of a given house. In other words it is a technical aid to a whole range of vital social and economic policies. On the one hand, the cost of auditing and rating individual houses is large and here there is room for collaboration between government, the fuel industries, private and public landlords, owner-occupiers and local authorities, all of whom stand to gain from energy rating. On the other hand, if intervention was initially limited to houses of very low rating (NHER≤2) the establishment of this point on the scale is a simple procedure and therefore less costly, but still requires most of the stock to be audited. This is basically what was proposed in the recently defeated Energy Conservation Bill 1994.

**1.5
Some
definitions**

The title of the Working Group contains two terms – 'affordable' and 'warmth' – which are defined in detail in Chapter 2. Although, of course, there is room for debate about these, the principles underlying the policies recommended by the Group are not dependent on acceptance of the Group's specific definitions.

**1.6
Possible
solutions**

The Group recognizes that there are both capital and revenue constraints on expenditure by government, the fuel industries and the private sector. Nevertheless the magnitude and urgency of the problem are such that substantial expenditure will be needed to remedy the situation in a reasonable timespan – say 16 years. Some measures, such as heating systems, have a limited lifetime and involve periodic expenditure. However, if savings on health cost, property values and indirect costs are taken into account, the net expenditure may be smaller or even negative (i.e. overall savings). The recommendations in section 1.7 try to balance the size and urgency of the problem with the economic constraints. In arriving at them the Group asked itself a series of questions, the answers to each of which can be located somewhere on a bi-polar scale. The recommendations emerged from the attempt to answer these questions, which fall into five groups of more general questions. Within each group there is a kind of critical path insofar as the answer to the first in each group determines the answers to the remainder.

The five general questions are:

1. **Why** is action necessary?
2. **What** kind of action?
3. **When** should action be taken and **who** should be included?
4. **How** are objectives to be achieved?
5. **How** much should be spent?

1.6.1 WHY?

(a) Should the policy on CO_2 emissions reduction be balanced against other objectives?

Chapter 2 identifies the evidence for the size of the problem. Chapter 4 looks at the effects of the energy used in domestic equipment on CO_2 emissions.

'Affordable' and 'warmth' have been operationally defined. Apart from savings in health, and in direct material and social costs, a policy that clearly benefits low-income families by enabling them to achieve affordable warmth has common-sense ethical values about the quality of life at its core. It may result in relatively small reductions in CO_2 emissions in the short term but this will be an intermediary state until such time as the present housing stock has been significantly improved. There is a hierarchy of solutions with consequential phased work:

1. First alleviate fuel poverty and associated effects such as ill health.
2. Next, proceed to improve energy efficiency.
3. Finally, reduce CO_2 emissions.

The interaction of these phases with government policy is explored in Chapter 5.

1.6.2 WHAT?

(a) What standards of 'warmth' should be adopted: full compliance or some staged compliance?

The target standards of 'warmth' have been discussed. To make these 'affordable' is a function of the thermal property of the house and its heating system; it is those which bring these issues together. The objectives for low-income households cannot be met unless the performance of their houses is raised to NHER≥8. The need to target limited resources implies a phased approach with a limit on the number of interventions in the same house. The remainder of the stock, where more fuel can be purchased, can be raised to that standard, perhaps in stages, at a later date. This policy brings together the social issues of income with the physical one of the condition of the housing stock, as discussed in Chapter 6.

(b) Should policies be related to climatic zones?

It is not only external air temperatures, on which degree-days are based, that affect fuel consumption. Wind and sunshine availability are also significant. An overall **climatic severity index (CSI)** can take account of these and its use shows that there can be differences as large as 60% in the fuel consumed to achieve the same standard of warmth in the same house located in two different climatic zones within the UK. Moreover, differences occur not only on a macro scale; microclimatic variations within a region or a site also cause significant differences. This is partly taken into account by using a rating system such as NHER which, unlike some others, is sensitive to climatic variations (but not yet to the CSI). The fuel subsidies required for those unable to afford warmth also have to be weighted by an appropriate climatic severity index.

1.6.3 WHEN AND WHO?

(a) What should be the time-scale for action?

Chapter 6 presents the argument that a time-scale of 16 years will be needed to raise the standard of all houses occupied by low-income households (defined as those in receipt of one or more of the 'passport' benefits used in the

HEES scheme) to NHER≥8, as the only way of providing them with affordable warmth. But the programme extends beyond this, as periodic visits will be needed to monitor that standards are being maintained.

(b) Is action proposed for the 'poor' (those on a means-tested benefit) or the 'nearly poor'?

The Group's focus is on affordable warmth for the poor. Those households on benefit are easily identifiable, but equity demands that the same approach will be needed for the 'nearly poor'. The evidence for this is presented in Chapter 2 and the argument in Chapter 6.

In combination the answers to questions (a) and (b) also answer the next four questions.

(c) Should intervention be incremental or total?

This question has several aspects. For individual houses, there is the option of one or a few measures at a time or an entire 'basket'. For streets, areas, estates and other housing blocks there is the option of carrying out one measure, such as cavity-fill, for all the houses at a time or tackling individual houses on the basis of economic or social criteria and therefore returning to the same area several times. The latter policy might attract social stigma and the take-up rate might be low. It would also cause repeated disruption. In both cases incremental work would spread a given resource further and have greater social equity. On the other hand it is more costly and causes greater individual household and area disturbance, and there are no economies of scale such as are obtained in the 'basket' approach, in estate treatment and in grouping work for a single measure. Local authorities and other agencies will be free to choose the most appropriate method in each case provided that the overall objective of NHER≥8 for the houses of all low-income households is reached within 16 years and provided that an upper limit is set for the number of operations in a single house.

(d) Should the work be triggered by the quality of the house or the income of the household?

Action should be targeted on the worst houses occupied by the poorest households. Poor quality houses occupied by better off households, or good quality housing occupied by poor people will therefore be excluded from the initial programme. Here too a flexible approach is needed to allow, for instance, marginal cases to be included as parts of a block so that economies of scale can be obtained. Such an approach will also make the eligibility/stigma issue less apparent and will mean that action is taken on the houses of some of the many households who are entitled to benefit but, for one reason or another, do not claim.

For privately rented houses the benefit to the landlord may be in the form of increased rent for letting houses which are more economical to heat. A subsidy which meets the difference between the extra value and cost may be needed.

(e) Should building legislation apply to new houses or be made retrospective?

Since the number of new houses added to the stock is a small percentage of the existing stock, the control of thermal standards through building regulations applying to new houses cannot achieve the 16-year target, even if they are made to apply to major refurbishment. There needs to be a mixture of policies to encompass the entire stock. Moreover, the heatability standard needs to be incorporated in the definition of the **Minimum Tolerable Standard** (Scotland) and the **Fitness Standard** (England and Wales). Currently, the requirement is merely for the presence of a means of heating, not that it be economic.

(f) To what extent can advice without investment achieve the objectives?

The houses which are the focus of this report are generally occupied by households with no capital resources. Therefore no amount of advice, without capital grants, will achieve the objectives. Nevertheless, discussion with occupants, or their landlords, has to be an important part of any strategy so that informed choices are made about the most suitable measures on building fabric, heating systems and energy-consuming domestic equipment. Continuing advice on the use of systems and installed equipment needs to be made available. For households where some resources for investment are available it has been found that marketing, information and education do achieve results, and there has been substantial investment by owner-occupiers in energy efficiency measures in the last two decades. Clearly such policies must continue.

1.6.4 HOW?

(a) What is the role of local authorities?

Local authorities are the only agencies having some responsibility for the entire housing stock in their areas. Therefore they are the only ones able to play a comprehensive role in auditing the stock and overseeing, monitoring and directing schemes of upgrading. This does not necessarily mean that they have to administer funds or carry out the work. Chapter 3 provides the evidence for their role. A local authority's progress towards reaching stated targets will need to be monitored by central government, which will require penalties or sanctions for making sure that targets are met. Chapter 7 explores how such schemes might work.

(b) What is the role of energy rating?

Initially a simplified audit of all houses has to be carried out in order to identify the worst-performing ones (NHER≤2). These data must then be related to those on household income and benefits, so that the houses for action are quickly identified. This can be done at very low cost by adopting an address-specific stock standard, say of Level 0 to Level 0 plus (both of which are explained in Chapter 7 but whose accuracy is not yet known). The programme from then on can have a good measure of flexibility. The database which records this information must then be upgraded and made more accurate as work proceeds and ratings are undertaken by households (Chapter 7). Equally, the emerging practice of energy labelling of domestic appliances should be accelerated.

(c) Should ratings be used as a way of identifying stock to be acted on or as a way of guiding action?

Initially, ratings are a means of identification and targeting early action. The same data can also be used to guide action at household level, but as investment will be subject to such factors as capital availability, occupant choice and condition of the house, it will rarely be possible to follow the most cost-effective approach indicated by the audit and computed according to the methods described in Chapter 8.

(d) Should ratings be carried out on the whole stock or only part of it?

This is a question of time. It has already been indicated that the basic rating should be carried out quickly for the entire stock so as to identify the houses, linked to data on households, for priority action. Once action on these houses is in hand, a start can be made on the more accurate auditing of the entire stock.

1.6.5 HOW MUCH?

(a) What total resources are required?

Chapter 4 explores the costs and energy efficiencies of domestic appliances, including the methods for heating domestic hot water.

For space heating, there are two components: capital for upgrading and revenue for income support for the fuel poor, until such time as the houses of these households have been upgraded to the standard that enables them to afford warmth. The rate at which the task can be tackled is constrained both by capital availability and by limitations on skilled labour, training and administrative resources. The capital needed has been estimated at £1.25

billion per annum for 16 years to get to 8 million households. The methods for arriving at capital investment estimates are indicated in Chapter 8. It is accepted that the 8 million low-income households require additional income for fuel until such time as their houses score NHER≥8. The initial cost of this depends on the level of support. At £10 per week it equals £4 billion per annum; at £5 it is £2 billion and at £1 it is £400 million. Whichever figure is chosen, it will fall – eventually (in theory) to zero – as the investment programme proceeds.

The possible sources of capital include the E factor, Fossil Fuel Levy, additional money resulting from the review of the Distribution Price Control in 1995, local authority right-to-buy accumulated capital, EU funds and revenue raised from VAT on domestic energy. The application of some of these to investment in low-income housing would require major changes of policy. The question of the administration of these funds is addressed in Chapter 7; and of the methods of calculating their energy and fuel effectiveness, together with the financial methodology for combining capital and revenue expenditure, in Chapter 8.

(b) Should available resources be used on capital investment or in income support for fuel?

The answers to question (a) outlines the Group's recommendations on this.

(c) What economic criteria should guide investment decisions?

Many energy efficiency improvements can be shown to be cost effective on the normal criteria for investment, even only taking into account energy savings (though the investor and the recipient of the savings is not always the same person). These include fuel switching and the installation of more efficient appliances (Chapter 4). Others are not cost effective but if other benefits, for instance health or social costs, or benefits to the building stock, are taken into account then they are cost-beneficial. Cost-benefit analyses of this kind are notoriously difficult to carry out but are the essential tool for bringing together all the economic factors. Whilst the monetary effects of the costs of borrowing, effects on rents and economies of scale would be taken into account, the criteria must be based on all the calculable social and economic benefits. Currently the transfers which such studies imply cannot be accommodated between government departments, primarily the Treasury and those of Health, Environment, Education and Social Security.

A cost-benefit analysis may point to the advantages of accelerating the role of demolition of the worst housing as any policy on energy upgrading must have regard to the general condition of the fabric. It may also highlight the benefit of intervening to ensure that existing programmes of refurbishment result in homes with NHER ratings ≥8.

(d) Should there be investment in projects that cannot be shown to be 'economic' by any criteria?

Some measures (for instance, the external insulation of solid walls) are clearly not economic, even with wider cost-benefit analysis, but are nevertheless essential for the welfare of the household. This is especially the case where the highest standards are to be achieved because the household income is most constrained – e.g. in the case of single, unemployed people under 25 years old. Short of providing adequate housing into which such households can be moved, there must in these cases be a measure of social subsidy on the capital, coupled (if need be) to additional income for fuel.

(e) Should the resources or energy improvements be regarded as part of the total housing programme or a separate policy requiring its own resources?

This is linked to the next question. If the resources come from the fixed capital for HIP, then they will compete with other essential needs – homelessness and getting rid of 'unfit' stock. The question indicates the need to reconsider the basis of the present housing investment system so that it includes both housing and energy efficiency objectives. It would be essential that hypothecated sums for specific targets be defined within an overall budget. Penalties and rewards, geared to the achievement of targets by local authorities, would be part of such a policy.

(f) Should policies be focused on the landlord or the tenant?

In principle, rents and recognized values for hard-to-heat houses should be lower than those of energy efficient ones. Appropriate criteria, based on property values, for assessing the percentage of grant available to a landlord will have to be developed. These will give levels up to 100% where landlords are able to show that they cannot recover any of the costs from reasonable rent increases. Otherwise grants will bridge the difference between costs and the extra loan which increases in rent can service. Where a rent policy can be exercised – for instance where Housing Benefit is payable – there also needs to be a linkage between the level of rent and the energy rating of a house as an inducement to landlords to carry out energy improvements. In the case of owner-occupier households in receipt of benefit any subsidy for improvements needs to be tapered off at no more than 40p in the £1, on a means-tested basis, as the income rises above the basic level, so as not to perpetuate the poverty trap.

1.7
Policy
recommendations   The UK situation with regard to affordable warmth is so starkly different from that in other comparable developed countries, and has become so deeply embedded in our housing tradition, that it takes an effort of the imagination

to free oneself from conventional assumptions and grasp its severity as a social phenomenon. It is quite clear that to remedy the situation in a reasonable period – and the Group has assumed a target to 16 years – whatever policies are adopted, will involve a substantial expenditure. This is to be set against the present recurring revenue expenditure resulting from inaction, for social security costs, health service use and the cost of other related services. In other words it is the cost effectiveness for **society**, in the short term, which is at stake, with benefits such as employment increasing in the long term. The sources for the expenditure will have to be a mixture of government, local authorities, the private sector, the fuel industries, earmarked taxation and owner-occupiers. EU funding should also form a central part of proposals.

With so many interactive questions, and such a range of possible answers, it is clearly impossible for the Group to recommend a single, hard and fast strategy. Instead, it has established and answered the kinds of questions that need to be addressed; the answers suggest the basic principles upon which any strategy should be based:

1. The Group rejects the need to choose between the use of resources to make energy improvements by capital investment and their use to subsidize the incomes of those in hardship. The former choice would leave large, but diminishing, numbers in hardship until the completion of the investment programme. The latter would perpetuate the existing situation and represent a never-ending drain on resoures. The strategy has to be a balanced combination of the two. The total can be represented as a fixed annual capital expenditure and a decreasing revenue expenditure which tapers off (theroretically to zero) as the cumulative capital expenditure increases.

2. As a matter of high priority, government funding should be made available for a major cost-benefit study of fuel poverty, the boundaries of which are drawn to encompass all the factors discussed in section 1.2 above.

3. Sixteen years should be adopted as the target date for thermal upgrading to an acceptable standard – ≥8 on the NHER scale – of the entire housing stock occupied by low-income households.

4. Until such time as this target is achieved, households have to be helped with fuel bills through a system which takes into account: the household needs (size, age, special heating needs etc.); household income (using initially the five 'passport' benefits which qualify for HEES action); the climatic severity of the address; the energy properties of the house and its heating system (using the NHER or other approved rating system); and the cost of available fuel. The system proposed by the Campaign for Cold Weather Credits forms a suitable basis for immediate emergency action, but would be further refined as it takes no account of the energy properties of the house, fuel costs or special heating needs. The Group proposes the Housing Benefit system as the most suitable vehicle for delivery.

5. Upgrading and additional income for fuel will not, initially, result in large reductions of fuel use. The conflict with policies of CO_2 reduction has to be resolved in favour of upgrading and additional income support, with the CO_2 effects countered by other policies aimed, in part, at non-domestic users and, in part, at those domestic users who have sufficient resources both to buy the amount of fuel they need and to invest some resources, as tenants or owners, in energy improvements. If it is assumed that adequate additional income for warmth is made available, then any improvements to houses will yield immediate fuel savings and CO_2 reductions. If this is not assumed, then the fact that improvements to the houses of low-income households will yield little in the way of fuel savings and CO_2 reductions in the short term should be viewed alongside the urgent need to improve living conditions. In the medium and long term they, together with all houses, will make a significant contribution to these reductions. Investment in domestic appliances that are more energy efficient will have immediate economic benefits to households as well as reducing CO_2 emissions.

6. Local authorities should be regarded as having responsibilities for, and knowledge of, the entire housing stock in their areas, not only that which they own and manage. Their responsibilities should cover energy rating, determining and possibly administering capital grants, providing advice to other agencies on priorities and information to the Housing Benefit offices or other agencies used as part of the system for subsidizing fuel expenditure.

7. Capital programmes should be based on priorities. Initially the houses with the lowest NHER rating should be selected (≤ 2). Moreover, it would be acceptable to tackle all improvements in a few stages even though the final total cost might thus be increased. The measures would be ranked according to cost-effectiveness criteria; for instance draughtproofing, insulation and new heating systems. Initially temperatures of, say 18 °C and 16 °C in winter for the two zones within the home, defined in Chapter 2 – the kind of temperatures to be expected in a house of NHER = ca. 5 occupied by a low-income household – may be accepted, eventually to be upgraded to 21 °C and 18 °C. In every case, such a strategy will require careful assessment of the most cost-effective investment and of the type of improvement which most easily lends itself to incremental treatment – e.g. improvements to heating systems.

8. Energy rating should form a legal requirement in all house transactions including change of ownership, change of tenancy and setting of rents. A target date needs to be set by when all owner-occupiers and landlords will be required to have their properties rated. The training of auditors and raters and the execution of ratings will require government or local authority resources. Assistance with individual ratings should be means tested, on the basis of the five 'passport' benefits (section 2.1.1).

9. The thermal requirements of the Building Regulations should be progressively made retrospective. The Group suggests 16 years. The

Minimum Tolerable Standard and the Fitness Standard should also be amended to include a requirement for heatability to defined temperatures at a maximum permitted cost.

Defining the problem 2

Sonja Hunt and Brenda Boardman

The general perception of life in Britain is that we have increasing affluence, higher standards of living and access to more energy-using equipment. At the same time, the furore that greeted the imposition of value added tax (VAT) demonstrated that many households consider they spend too much on fuel already and could not afford extra expenditure. This paper looks briefly at the background to these two scenarios to see whether both could be true and, in particular: are standards of warmth rising equally for all income groups, or only for the better off?

The underlying reason for the analysis in this report is concern that the ill health and hardship caused by fuel poverty have been seriously underestimated, with serious implications both for the individual sufferer and for society.

2.1 Domestic energy consumption

Trends in domestic energy consumption can be seen to result, at least partly, from the socio-demographic changes which have occurred over the last 30 years. The information used comes from several sources so the geographical coverage and time periods vary.

- The population of the UK was 58 million in 1992 and is increasing slightly. Over the past ten years, the increase has been approximately 0.2% per annum and this is expected to rise to 0.3% per annum for the period 1990–2010.
- People are living in smaller groups. Average household size has decreased from 2.9 persons per household (pph) in 1971 to 2.5 pph in 1991 in Great Britain; a major influence has been the doubling of one-person households between 1961 and 1991. Government forecasts are that household size will continue to drop, but more slowly, to 2.4 pph by 2000. Beyond that, household size is expected to remain relatively static, though this is difficult to predict. Household size in the Scandinavian countries is already smaller.
- The number of UK households was 23 million in 1992 and, because of the trend to fewer people per household, is increasing faster than population size, at an average rate of 1% per annum since 1970. This trend is also expected to slow down over the next 20 years but there

will still be some growth in household numbers.

- The floor space occupied by each household has been decreasing over the long term, as a result partly of smaller new construction and partly of the conversion of existing large dwellings into several separate units.
- In the more recent past, since 1980, the average floor area per household in England has been static at 80 m². There are no predictions about space trends, although there is a tendency for new dwellings to be smaller than the present average.
- Fewer people (per household) are occupying a constant amount of space so that the average amount of space occupied per capita is increasing. For example, in England, between 1980 and 1990 each person had an additional 2 m² of floor space. The average is now 32 m² per person.

There is a complex relationship between household energy use, space occupied and the number of people in the household, with a relatively small drop in consumption per household as households contain fewer people (Figure 2.1). For instance, there are economies resulting from the need for less hot water (baths) but each household still needs a refrigerator and the same number of rooms to be heated and lit. The Dutch data cannot be replicated for the UK, but the principle is the same and is confirmed by the trends: energy consumption per capita is increasing, whereas average household energy consumption is dropping slightly. This means, that total energy demand will increase if the same number of people live in smaller groups.

These demographic factors are important as they are working in opposition to trends in improved energy efficiency.

The average UK household now owns more electrical appliances (Chapter

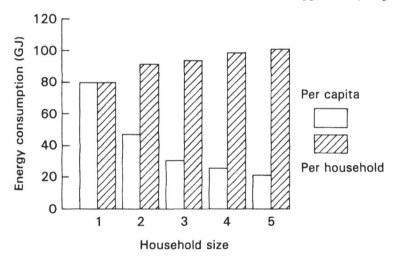

Figure 2.1 Household size and energy use, The Netherlands, 1985 (source: Schipper and Meyers, 1992).

4), though many of the more recent acquisitions, such as video-recorders and computers, have a relatively low demand for electricity. The Building Reseach Establishment has used the data on household size, appliance acquisition, energy efficiency improvements and energy purchases and calculated the likely resultant temperatures. They have assumed a constant differential of 2.5 °C between the temperatures in centrally heated and non-centrally heated homes and show an average increase, across the whole housing stock, from 12.79 °C in 1970 to 16.14 °C in 1989. At these rates of increase, comfort level 'could be reached in perhaps 30 years time' (Shorrock *et al.* 1992).

Unfortunately, there are few surveys of actual measured temperatures inside the home and those that exist are difficult to compare. Therefore, only the most generalized temperature trends can be identified (evidence of cold homes is given below).

Over the period 1970–86, total energy demand in the domestic sector had been relatively constant, when external weather variations were taken into account. Since then, demand has been increasing and the continuing decline in household size will place continuing pressure on growth in demand in the next few years. With total domestic energy demand now growing, the implication is that the effect of more efficient appliances, heating systems and greater insulation levels in dwellings has not been sufficient to offset the effect of demographic factors, increased energy services (such as greater warmth) and higher levels of appliance ownership.

**2.2
Energy use
and carbon
dioxide**

The costs of the main domestic fuels vary, as do the amounts of carbon dioxide emitted per unit of delivered energy. Using UK averages, in 1989, the greatest range is between on-peak electricity and gas, which vary by up to a factor of four in both carbon dioxide and price per unit of delivered energy (1). The range between the fuels is less when measured in terms of useful energy services, such as warmth, because the efficiency of the appliance has to be taken into account. However, it is as delivered energy that we purchase fuel. As a result, it is important to clarify the units being used when discussing the breakdown of energy use in the home (Table 2.1). For instance, in the average home, space heating uses 61% of the energy but only represents 40% of the costs.

The carbon dioxide emissions from electricity have been reduced over the last few years by the increased use of nuclear power. The construction of gas-fired electricity generating capacity (the 'dash for gas') and the displacement of coal-fired stations will continue to reduce the emissions from electricity. Both of these trends have already been incorporated into the Government's projections of carbon dioxide emissions until the year 2000. Thus, compliance with our Rio commitment to curb emissions even further and to reduce output of carbon dioxide by 10 MtC will depend, in all sectors, on reduced energy consumption.

Table 2.1 UK domestic energy consumption, CO_2 emissions and fuel expenditure, by use 1989 (Source: BRE, NCC)

	Energy	CO₂ emissions	Fuel expenditure
Space heating	61	47	40
Water heating	24	21	18
Cooking	05	07	07
Lights and appliances	10	25	25
Standing charges			10
Total	100%	100%	100%

Ever since the first oil crisis in 1974, the Government has been urging the use of less energy, or to use it more efficiently (Chapter 4). The expenditure that has been incurred under various programmes is outlined in Chapter 3, but has been insufficient to curb consumption in the domestic sector.

2.3
Fuel poverty   **Fuel poverty** is the inability to afford adequate warmth because of the energy inefficiency of the dwelling (which the Group takes to include the fabric, heat emitters and appliances). To overcome fuel poverty, households have to be able to obtain affordable warmth in the house that they occupy, on their present income. The problem has been created by a lack of capital investment in the energy efficiency of the dwelling and will only be ameliorated, over the long term, by the provision of energy efficient dwellings. This situation is aggravated by poverty, because a higher level of energy efficiency in low-income homes is required if warmth is to be affordable on a low income. However, the problem of fuel poverty has been created by low levels of capital investment, rather than just poverty or high fuel prices (as the name implies).

Fuel poverty is a consequence of housing rather than climate as is indicated by the fact that the problem does not occur in other countries which have much colder and more inclement weather – for example, Canada, Denmark, Sweden, Norway, Finland, northern Germany and the northern USA.

The achieving of **affordable warmth** indicates that fuel poverty has been overcome, although it only refers to heating, the main emphasis of this report. However, it is acknowledged that there will often be a shortage of other energy services in cold homes such as insufficient hot water and poor lighting conditions. Expenditure on these other services also has to be 'affordable', though the standards that should be achieved have been poorly researched and are difficult to define. For instance, how many baths a week represent an appropriate standard for cleanliness and well-being? How much hot water is needed to wash clothes, whether by hand or in a washing machine? Chapter 4 develops these issues further, but more research is needed to clarify the necessary expenditure on these other energy services. For the purposes of this report, the assumptions built into the energy audits currently in use are assumed to be appropriate. This assumption is by default rather than as a result of conviction.

Energy inefficiency arises from a lack of sufficient insulation and inappropriate ventilation, combined with a heating system which is inefficient or uses an expensive fuel and is thus too expensive for the occupants. British legislation still does not require the provision of an adequate and affordable heating system in a dwelling, either when new or as part of the definition of fitness for human habitation. In the latter case, in England and Wales, the 'system' can be solely an electric socket. The situation in Scotland is worse.

A major problem faced by the Working Group has been insufficient data on the housing conditions of the poorest households. Several surveys show that low-income families are less likely to have insulation measures, the cheapest fuels or the most efficient heating systems. The homes of the poor are more expensive to keep warm than those occupied by better-off households. The difficulty has been in quantifying the level and extent of deprivation. For instance, it is not known how many households have properties that would be rated below 2 on the NHER scale (discussed in Chapter 3), although 20% of the Scottish housing stock is below this figure.

2.3.1 NUMBERS AFFECTED

The Government has stated (*Hansard* 1.4.93, WA col. 492) that in Britain today there are 8 million households in receipt of one of the means-tested benefits (Income Support, Housing Benefit, Family Credit, Council Tax Rebate) or Disability Living and Disability Working Allowance. These six benefits act as a 'passport' for the original Home Energy Efficiency Scheme (**HEES** has been extended to all those over 60, whatever their income, in the the VAT compensation package). These 'passported' households have been defined as **poor** by the Government and therefore in need of assistance with their weekly bills. By definition, these 8 million households also have little or no capital and so are unable to fund even minor improvements themselves.

In addition to existing claimants, there are large numbers of people who are eligible for means-tested benefits but do not claim them. Non-claimants could represent another 1 million households, bringing the total in poverty to 9 million households.

Finally, there is a category that could be called the **'nearly poor'** – those households that just fail to qualify for a means-tested benefit, but who are also unable to afford adequate warmth to a considerable degree. This group cannot be quantified at all because, for fuel poverty purposes, there is an interaction between income and the energy efficiency of the dwelling. In properties that are extremely poorly insulated and expensive to heat, a higher level of income is necessary to obtain 'affordable' warmth than in a more efficient property. Only the very rich could afford to be adequately warm in the most inefficient properties, without capital investment.

The numbers in poverty are growing: there was a 14% increase in income support recipients in one year from 4.66 million in 1991 to 5.32 million in 1992. Furthermore, the real income of the poorest 20% of households fell by 3% between 1979 and 1991, whereas the real income of the richest 20% rose by 49%. The disparities of income are growing in the UK and poverty is becoming both deeper and more widespread, whatever definition is used.

It is exceptionally difficult to judge the effect of these different trends and measurement methods. The Group has accepted the Government's statement that 8 million households are eligible for assistance. Some of these households may be living in good quality accommodation that does provide them with affordable warmth, though this is probably only about 100,000 households. The Group uses the 8 million figure as the number of households suffering from fuel poverty. This is approximately 36% of all UK households. For statistical convenience, some of the data in this Report take 30% (7 million) of all UK households as the necessary definition of low income. This is only for the purposes of analysis. The policies are targeted on 8 million households, as a minimum.

About two-thirds of the households with the lowest incomes live in rented accommodation (private, local authority, housing association or rent-free) with no legal obligation to improve the fabric of their homes (the numbers are given in Table 3.1, Chapter 3). The majority of the remaining households are elderly people who own their homes outright. Some of the latter households do have small amounts of capital – usually less than £600 – which they reserve as a funeral fund. Thus at least 8 million households are in need of outside assistance to implement energy efficiency improvements.

2.4 Affordable warmth

The title of the Working Group contains two terms – 'affordable' and 'warmth' – which need to be defined before possible solutions can be discussed.

2.4.1 WARMTH

'Warmth' is a difficult concept to define. It describes conditions of comfort. For protection against cold-induced illness the temperature threshold is lower, and to prevent condensation and mould (and the resultant allergies) the thresholds are lower still. Levels of heat in a home should be such as to provide for the comfort of the occupants and the maintenance and protection of their health. It is generally agreed that the mean temperature required to provide comfort for people at home and awake is 21 °C.

It is not easy to establish minimum temperatures for health in the absence of strict scientific proof (which could only be obtained by unethical experiments). However, evidence exists to support the contention that the risk of respiratory impairment increases below 16 °C, that cardiovascular strain occurs below 12 °C and that the risk of hypothermia increases as temperatures fall below 6 °C. The World Health Organisation recommends a minimum air temperature for the sick, the handicapped, the very old and the very young of 20 °C. Any dwelling should be capable of sustaining these temperatures throughout the year regardless of the outside climate in all rooms as necessary.

Conditions for comfortable warmth depend on four physical variables and two personal ones. The four physical ones are air temperature; the radiation exchange between the body and surrounding surfaces; air speed; and humidity. The two personal ones are the amount of clothing and the level of activity. An infinite set of combinations of these six yield the same subjective sensation of warmth. For instance, in low air temperatures, warm radiating surfaces can compensate; similarly, even though air temperature may meet normal standards of warmth, the presence of large, cold surfaces may produce a sensation of cold. The Group, whilst recognizing the complexity of thermal comfort, has assumed that surfaces such as walls, floors and windows are no more than a few degrees below air temperature (a condition not difficult to attain with reasonable insulation and double glazing), that there are no strong draughts (a condition that will apply with draught-proofed doors and windows and controlled ventilation), and that the relative humidity (a variable of little significance in temperate conditions) is below 70%.

For clothing, normal indoor clothing in living spaces and adequate bed covers in bedrooms are assumed; and for activity, a sedentary occupation for living rooms and a slightly more active one for other areas. That justifies slightly higher temperatures in the former. For bedrooms, when used for sleeping, slightly lower temperatures are needed than for living rooms, as normal bedclothes have a higher insulation value than indoor clothing. It is recognized that bedrooms may be used for other activities than sleeping, which should mean that they are heated to the same temperatures as living rooms. In practice, however, the Group has accepted the convention of lower temperatures for bedrooms even for daytime use and assumed a shorter period of occupancy of five hours during the day. Whilst somewhat arbitrary, this does allow for uses such as afternoon naps and evening homework.

People are in the home for varying numbers of hours but follow remarkably similar activity patterns: over 50% of all time is spent in sedentary activity, resulting in low metabolic rates. Less than 20% of time is moderately active, for instance making the bed, sweeping the floor (Boardman, 1991, pp. 106–109). The majority of low-income housholds are in the house, rather than out of it, for most of the day, because they are pensioners, unemployed, looking after young children, sick or disabled. Typically, they occupy the home for 13 hours during the day. Excursions from the house mean that they are out for 3 hours in total, leaving an average of 8 hours when they are in the house and asleep.

Whilst average conditions of 'warmth' can be aimed for, in any population there will be variations around this average due not only to inter-personal differences but also to the specific needs of the elderly, the sick and disabled (who may be relatively immobile) and very young infants.

On these assumptions it is acceptable to define 'warmth' by means of air temperature. The target temperatures adopted are 21 °C for living rooms and 18 °C for other spaces, giving a mean house temperature, when the house is occupied, of about 19–20 °C. It is fully recognized that these are not only above those generally found, but significantly so in the case of poor households. They represent conditions for comfort, for health and for the avoidance of condensation. The rationale behind aiming for comfort is that the issues discussed in this report, whilst of an urgent nature, lead to policies which have a long time horizon.

It is not sufficient to define target temperatures for various spaces in a dwelling. The duration for which these should be met also has to be defined in terms of hours of the day, days of the week and months of the year. The Group has assumed that these targets apply to the entire dwelling for the whole year. Since occupancy varies (for instance, between a household whose members are all at work or school and one composed of pensioners and others 'at home'), fuel requirements for heating as well as other uses vary correspondingly. The following times, therefore, whilst notional, should be achievable by all low-income households:

- Living rooms: 21 °C for 13 hours a day.
- Bedrooms: 18 °C for 13 hours a day (including 8 hours at night).
- Other spaces: 18 °C for 13 hours a day.
- All spaces: 14.5 °C at all other times.

Such a temperature regime is likely to result in a temperature, averaged over space and time, of about 17 °C. Evidence of lower temperatures therefore indicates cold homes, for the purposes of this study.

There are millions of homes in Britain where this minimum temperature is not achieved. As Figure 2.2 indicates, 20% of homes in England have a temperature below 12 °C when it begins to freeze outside and only 50% have a mean internal temperature over 16 °C. The number of cold homes increases as the weather gets colder, indicating that many households have limited amounts of money for heating and are unable to increase their expenditure to compensate for a drop in external temperatures. Furthermore, the strain of heating an inefficient property in severely cold weather may be causing the household to cut back on food, at a time when to do so causes additional health risks.

2.4.2 AFFORDABLE

The definition of 'affordable' is based on current expenditure on fuel by low income households. In recent years, the 30% of households with the lowest incomes have spent around 10% of their weekly budget on fuel (including standing charges). The actual percentage has varied as a result of

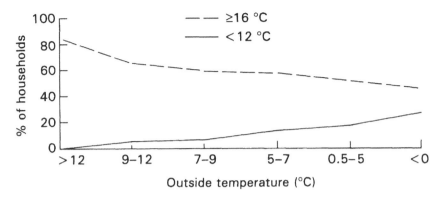

Figure 2.2 The effect of outside temperature on mean internal temperature. (Source: EHCS 1986, Supplementary Energy Report.)

fuel prices, increasing poverty and the weather, though the effect of these different parameters has not been identified. In all years, the poorest households spent less money than richer ones, but over twice as much as a proportion of expenditure. This indicates the importance attached to warmth and other energy services by even the poorest households (Table 2.2).

In 1988, among low-income households, single parents allocated an average of 15.8% of their budget to fuel, and pensioners 14.4%. The average household expenditure for the country as a whole was 4.7% of income. Thus, people on the lowest incomes need to spend a disproportionate amount of their income, more than twice the national average, on fuel. The average low-income household is assumed to be able to afford 10% of expenditure on fuel, or just 6% if only space heating is considered. It is crucial to remember that, despite this emphasis in the weekly budget on fuel expenditure, the poor were still not able to obtain adequate warmth and had colder homes than the rich. These differentials are even more more marked in Scotland and the North of England where average outdoor temperatures are lower.

Table 2.2 Expenditure on fuel by UK households. (Source: *Family Spending*)

		30% of households with the lowest incomes	70% other
Weekly fuel expenditure	1991	£10.02	£13.21
% of weekly budget	1991	9	4
	1988	10	4
	1985	11	5

The importance of the proposals in this Report is to ensure that, for the same level of expenditure, the poorest households are warm. This cannot be achieved by additional income alone, as the poor occupy the least energy efficient houses: even with an extra £3.19 per week in 1991 (from Table 2.2), they would not have the same temperatures as achieved by better-off families. For instance, research for the government has estimated that the average low-income pensioner household needs to spend an extra £10 per week to obtain adequate warmth in their present local authority housing. This research was undertaken before the imposition of VAT, so that the figure from next April should be £11.75 per week at least. If this additional expenditure is typical of all low-income households (and this is not known) it would require a total extra annual expenditure of over £4 billion*, and result in the emission of an extra 13 MtC (as carbon dioxide) every year.

The present social security system does not provide any benefits to assist with exceptional heating costs on a regular basis. There were Heating Additions prior to 1988 going to over half of all claimants, but these have been abolished. The Social Fund will pay £7 per week to 2.6 million eligible households when the weather has been below freezing for seven consecutive days. In most years, this scheme is triggered in one or two weeks resulting in about £8 million expenditure – a miniscule amount in comparison with the £4000 million extra necessary for adequate warmth.

The VAT compensation package announced in November 1993 does provide additional income for pensioners and claimants for the next three years but the present scheme does not address subsequent years. Most of the extra £400 million will be given to the elderly, rather than the poor, and is based on average expenditure. There is no allowance for variations in the energy efficiency of the property. Therefore the money will help, but it will be insufficient to offset the extra costs, particularly in energy inefficient and poor homes.

The lack of affordable warmth has adverse health and social consequences and leads to financial costs for the individual, for local and government agencies and for the country as a whole.

2.5 Health and social effects of cold, damp housing

A considerable body of research carried out in the UK, Canada and Scandinavian countries has, in recent years, succeeded in disentangling the health effects of damp, cold housing from those due to associated factors such as poverty, overcrowding, unemployment, smoking, selection bias, reporting bias and other relevant confounding variables.

Damp and cold housing can have adverse effects on health in several ways:

1. By encouraging the growth and proliferation of **pathogens** in the indoor environment.
2. Through **physiological** changes brought about by the indoor climate.

* Here and throughout billion = thousand million.

3. By engendering **emotional distress** associated with discomfort, tension, frustration and despair.
4. Through long-term effects on the social, emotional, intellectual and physical development of **children**.

2.5.1 PATHOGENS

Damp conditions in a dwelling, especially as a result of condensation, encourage the growth of moulds which may be visible on walls, ceilings, carpets and clothing and present in the air as fungal spores. Several kinds of fungus found in houses, such as *Aspergillus*, *Cladosporium*, *Fusarium* and some varieties of *Penicillium*, are known to give rise to adverse effects of three kinds: allergies, infections and toxic reactions.

Exposure to fungal spores has been found to cause respiratory conditions such as allergic asthma and other forms of inflammatory disease of the respiratory organs, rhinitis, pneumonitis and alveolitis.

Mycoses arising from fungal spores are opportunistic infections which can invade the body tissues and which pose a health hazard for neonates, the elderly, patients undergoing immunosuppressive therapy, AIDS patients and other individuals whose immunity system may be compromised.

Toxic substances from domestic fungi can be taken up by inhalation and also, possibly, by ingestion of contaminated food or, in the case of small children, directly from hands which have touched mouldy patches. Certain moulds such as *Stachybotrysatra* can be particularly virulent and give rise to severe symptoms including lethargy, fever, headache, sore throat and aching joints, which disappear upon removal of the patient from the dwelling. Chronic fatigue, nausea and diarrhoea can also be a consequence of toxicosis.

At least two large-scale epidemiological studies have found dose response relationships between the presence of mould and symptoms of allergy, infection and toxicosis. These findings were independent of income level, smoking, unemployment, heating and washing arrangements, household size, selection or investigator and respondent bias. The evidence linking certain types of fungus in domestic dwellings to health problems is now such as to justify the assumption of a causal relationship.

Damp housing also encourages the proliferation of house dust mites which are known to precipitate Type 1 allergies such as asthma.

2.5.2 PHYSIOLOGICAL CHANGES

There is both epidemiological and experimental evidence to indicate relationships between cold indoor temperatures and physiological changes associated with both respiratory disorders and ischaemic heart disease. Cold air can trigger bronchospasm. Indoor temperatures below 16 °C increase the risk of respiratory infection, particularly in young children, and the risk of hypothermia in the elderly.

Daily deaths in England and Wales increase from 4.9 to 6.9 per million for myocardial infarction and from 3.2 to 4.8 for strokes when the daily temperature falls from a high of 17 °C to −5 °C. At least 25% of excess winter deaths occur in the community rather than hospitals. A drop of only 1 °C in the average winter temperature results in an extra 8000 deaths in that winter. It should be noted that excess winter deaths are much lower in countries such as Sweden and Finland, which have lower external temperatures but where building and heating standards are higher than in Britain and where households can attain affordable warmth.

Two risk factors for heart disease – hypertension and elevated fibrinogen levels – are related to low temperatures. Seasonal variations in blood pressure occur which have been clearly related to indoor temperatures. Data from studies in Finland indicate that the decline in heart disease in that country is closely associated with improvements made to the housing stock such as adequate affordable heating, increased insulation and appropriate ventilation.

2.5.3 EMOTIONAL DISTRESS

A number of studies using structured methods and controlling for other relevant factors have found damp and cold housing to be a cause of emotional distress in women and children, giving rise to anxiety, depression, headaches and fatigue in adults and irritability, temper tantrums, bedwetting and headaches in children.

2.5.4 CHILDREN: LONG-TERM EFFECTS

There is strong evidence to support the contention that children who are often subject to ill health are likely to grow into adults who are vulnerable to a variety of health problems. Such children miss time from school and social and play activities with other children. Where the housing conditions are such that a room or rooms cannot be used, children may have to share the parental bedroom or try to do their homework in a room where many other activities are taking place. Lack of space and privacy are known to have deleterious social and emotional effects. Thus, particularly for children there can be long-term consequences for physical and emotional health and for social, sexual and intellectual development.

2.5.5 VULNERABLE GROUPS

Since there are at least 8 million households living in conditions of fuel poverty, a high proportion of the members of these households are at risk of health problems which are a direct or indirect result of housing conditions. Certain groups are more at risk than others: for example, neonates and children under the age of 4 years, elderly people on low incomes, unemployed men and women under the age of 25 on income support, the physically and mentally disabled and those who are chronically ill. These groups are vulnerable on one or more of several counts: low income, impaired mobility, spending much of the day at home and/or the presence of a medical condition which lowers immunity.

Studies have shown that people who live in adverse housing conditions and have been hospitalized are more likely to have to return to hospital than those whose houses are more comfortable. This creates a 'revolving door' syndrome. Recovery from illness is delayed and the risk of relapse increased where the dwelling is cold and damp.

2.6 Financial costs

2.6.1 COSTS TO THE OCCUPANTS

The need to spend a high proportion of disposable income on heating may lead to economies on food and clothing which, inadvertently, may raise further health risks. If the tight budget results in rent arrears, then occupancy of the home is also at risk.

Fuel debt imposes a further strain on households when they are unable to meet the disproportionately large bills. Credit metering requires families to save a large amount of money over a three-month period. Where income is low, this may prove impossible to do in the light of other expenses. Although it is true that disconnections for fuel debt can be avoided by changing to easy payment and budget schemes, these payment methods bring other problems. For instance, with prepayment meters, the tariff and/or standing charge is higher, creating a further debilitating drain on the household in order to 'assist' with budgeting. (The utilities' explanation is that the prepayment meters and the dispensers are expensive to install.) Self-disconnection occurs when there is no more money for fuel until the next giro or payment arrives. These 'hidden' disconnections conceal private suffering which often occurs on a regular basis, with a complete absence of fuel, perhaps for one, two or more days. The drop in disconnection statistics can, therefore, be hiding other increases in fuel poverty.

The collection of benefits, in itself, is often seen to entail a stigma, and may be one of the reasons why so many fail to claim their entitlement. Providing futher benefits to offset, for example, the imposition of VAT on domestic fuel, though an essential stop-gap, increases unwanted dependency and has an adverse effect on self-esteem. The Group has proposed additional benefits only because any capital investment programme, to be effective, would take so long that help is needed in the interim to limit the amount of hardship.

Families also incur extra financial burdens both as a consequence of the housing conditions themselves (for example, frequent redecorating to hide mould contamination) and as a result of health problems. Time must be taken off work either for personal ill health or to look after sick children and, since many are low-paid workers with no job security, there is a risk of losing the job altogether. This may result in reluctance to miss work, compounding the health hazards.

The accumulation of arrears in fuel bills, and/or debts incurred in other areas of expenditure in an attempt to meet fuel bills, can lead to the need to borrow from moneylenders at excessive credit, further deepening the poverty trap.

Damp causing damage to furniture, furnishings, decorations, clothing and toys leads to further expense. Depending on tenure, costs of damage may fall either on the landlord or on the occupant. Where condensation and mould are rife the costs to a landlord of redecoration, voids or a high tenancy turnover mean that improving the energy efficiency of the dwelling will accrue greater savings to the landlord than the tenant.

2.6.2 COSTS TO SOCIAL AND NATIONAL AGENCIES

Damp and cold housing imposes considerable financial costs on the Health Service in both the short and long term. Families living in such conditions are more likely to consult their GP for both physical and emotional complaints. A child from a cold, damp home is more likely to end up in hospital, to stay there longer and to have to return. Respiratory, heart and cerebrovascular disorders make up the major part of all causes of mortality and morbidity, at least a portion of which can be attributed to housing conditions and is responsible for a large slice of the Health Service budget. It has been calculated that adverse housing conditions are responsible for an extra £300 per year per inhabitant in GP consultations, hospital and medication costs, with an avoidable expense to the NHS well in excess of £1000 million per annum.

The 1991 Department of Health document *Health of the Nation* targeted several medical conditions for which prevention and control are important. Among these are heart disease, strokes, childhood asthma and mental illness. The provision of affordable warmth in all households would do much to prevent and ameliorate such conditions, as well as relieving the current strain on primary, secondary and tertiary health care.

Family disruption and conflict are engendered by damp, cold housing. Family members may be reluctant to come home to uncomfortable conditions; available space shrinks in winter due to the inability to heat the whole dwelling and parents and children may have to use the same room for living and sleeping. Lack of adequate finances and uncomfortable living conditions lead to tension and quarrels. Some of the consequences of this may impose extra burdens on social services and voluntary bodies.

Damp and cold homes are unsuitable places for a child to work and school staff may find themselves trying to compensate for unsatisfactory homework conditions. Time missed from school needs to be made up and teachers will need to help with this.

The enlargement of a complex benefit system creates further bureaucracy and resulting administration costs to the taxpayer.

Local authorities and private landlords are faced with loss of rent and loss of available accommodation from unlet and unlettable houses.

Low levels of capital investment in the housing stock lead to the generation of conflict and sometimes lawsuits between, for example, landlords and tenants.

Failure to address the issue of affordable warmth also leads to continuing deterioration in the building fabric with a consequent increase in repair and maintenance costs to private landlords, local authorities and to owner-occupiers. There is a loss of market value and a decrease in habitable housing, possibly leading to increases in squatting and homelessness or the use of unsuitable habitations.

Figures published by the Building Research Establishment suggest that, without intervention, it will be about 30 years before the desirable temperatures for dwellings are met in the **average** house. In those occupied by the poorest sector of the community it will be longer, if indeed it ever happens at all.

2.7 Conclusions

Total domestic energy demand, when weather-corrected, is now growing. As the major opportunities for fuel switching in electricity generation have already been accounted for, the Group's commitment to curb carbon dioxide emissions can only be met through reduced energy consumption. Ever since the first oil crisis in 1974, the Government has been urging people to use less energy, or to use it more efficiently, but to little effect. Therefore, any reduction in demand will be difficult to achieve and require a new and dynamic set of policies.

Low-income households are different from the average: their homes are less energy efficient and more expensive to keep warm. The temperatures inside these poorer homes are consistently lower than those in the homes of better-off families, even though low-income households have to be in the house for more hours a day and should be recording warmer homes. Those that need the most warmth have the least money and have to buy an expensive product.

The poor are becoming poorer, certainly relative to the rest of the population, but often in absolute terms. This implies (though this is conjecture) that the temperatures achieved in the poorest homes may be declining, particularly as there has been minimal improvement in the thermal efficiency of these properties. The imposition of VAT will make even more households, both poor and not so poor, colder with the risk of a rise in morbidity all year and, particularly in winter, of mortality.

The lack of financial resources used to combat fuel poverty has resulted in substantial social costs: to landlords with unwanted properties; to the National Health Service for ill health from condensation, mould and cold; to the utilities with fuel debts and slow payers; to the poor who suffer so extensively. These social costs, which the Group includes in the phrase 'cost benefit analysis', are not taken into account in central government planning, though they would diminish significantly over the years if capital investment reduces fuel poverty.

For all these reasons, the Group believes that the 8 million households on the 'passport' benefits should be given sufficient income to afford to be warm. In the short term, until the capital investment can be undertaken this will result in additional carbon dioxide emissions. The contribution required by the rest of the population, at home, in travel and at work, will have to be commensurately greater if the UK is to meet its Rio commitments. However, the Group feels that the health and welfare of the present population has to be recognized as well as that of future generations.

Notes   (1) The exact comparison depends on stations needed to supply demand, which in turn depends on the weather. In addition, there is a smaller range in Scotland, because of hydro power, and consequently a higher range in England and Wales alone.

References   Boardman, B. (1991) *Fuel Poverty*. Belhaven Press, London.

BRE (1990) *Energy use in buildings and carbon dioxide emissions*. Shorrock, L. and Henderson, G. Building Research Establishment, Watford.

Department of the Environment (1991) *English House Condition Survey 1986, Supplementary (Energy) Report*, HMSO.

Department of Health (1991) *The Health of the Nation*.

Energy Efficiency Office (1990) *Energy Consumption Guide 2: A councillor's guide to affordable warmth for tenants*, Best Practice Programme, p.2.

Central Statistical Office (1993) *Family Spending: a report on the 1991 Family Expenditure Survey*, HMSO.

National Consumer Council (1993). *Paying the price*, HMSO.

Schipper and Meyers (1992) *Energy efficiency and human activity: past trends, future prospects*, Cambridge University Press.

Shorrock, L.D., Henderson, G. and Bown, J.H.F. (1992) *Domestic Energy Fact File*, Building Research Establishment, Watford, p.31.

Existing and likely initiatives 3

William Gillis

Expenditure on the nation's housing stock comes from a wide variety of sources. It is perhaps most useful to divide the stock into public and private sectors when examining funding sources. The **public sector** consists largely of property owned by local authorities, and housing associations. The **private sector** stock consists of owner-occupied and privately rented accommodation. The breakdown of the housing stock in tenure types is included in Table 3.1.

Over half of council tenants are claiming a means-tested benefit, as are a significant number of owner-occupiers. The smaller numbers in the private rented and housing association sectors nevertheless represent significant proportions of the households in these sectors.

The major sources of funding for capital investment in the local authority stock is the **Housing Investment Programme (HIP; Housing Resource Allocation in Scotland)**. Central government provides local

Table 3.1 Breakdown of UK Housing Stock by Tenure (source: BRE unpublished report, based on *Domestic Energy Fact File*)

	Numbers (000s)		Percentage of low-income households	
Tenure	Overall	Low income	All stock	Tenure group
Owner-occupied	14201	2556	11.98	18.00
Local authority	04972	3109	14.57	62.53
Private rented	01571	0693	03.25	44.11
Housing association	00596	0396	01.86	66.55
Total	21340	6754	31.65	31.65

authorities with borrowing approval for the necessary resources to fund their agreed HIPs. In addition to the HIP borrowing approval, local authorities are able to spend a proportion of the receipts, from the sale of housing and other assets, on their housing stock.

Local authorities are also able to fund repair and improvement work directly from their **Housing Revenue Account**. However, the pressure on the Housing Revenue Account in most local authorities is such that it is unlikely that anything more than the most urgent repairs would be funded in this way.

Capital funding for Housing Associations is provided via the Housing Corporations' **Approved Development Programme (ADP)**. The ADP provides loans and grants to associations to fund their capital programmes. Capital grants are not available for improvements and associations' own revenue resources are fully stretched in keeping new rents affordable and maintaining existing stock.

The resources identified above have to be stretched to cover a range of capital investment needs including major repair work, improvements and modernization and new build, although it should be noted that in recent years local authorities have been prevented by government funding restrictions from building new housing. It is important to note that improving the energy efficiency of the stock is only one of the many competing demands on these funds. The Audit Commission's analysis of the DoE enquiry into local authority stock conditions in 1985, for example, identified the total expenditure required on local authority housing in England and Wales as £10.8 billion for maintenance and £8 billion for improvements. At current rates of investment it will take more than 10 years to remove this backlog. This would be on top of normal routine repair and maintenance expenditure. Clearly there is severe competition for housing investment resources in the public sector.

In the private sector, the main source of funding for repairs and improvements to housing is the resources of individual owner-occupiers and private sector landlords. The Department of the Environment estimates that the annual expenditure on housing repair and improvement is approximately £15 billion. In the owner-occupied sector, the majority of investment takes place when a property is sold either to meet the requirements of lenders or to make the properties more attractive to prospective purchasers. Energy efficiency standards do not currently rank highly in the priorities of lenders or the majority of prospective purchasers. Owner-occupiers who suffer from fuel-poverty are unlikely to have the capital to invest in energy efficiency improvements to their properties and private sector landlords are unlikely to make significant investments in properties while the major benefits, in the form of reduced energy bills, will fall to their tenants. There are a number of means-tested grants available for low-income owner-occupiers and private sector landlords. **House Renovation Grants** are available for major repair and improvement work in England, Wales and Scotland and

Minor Works Assistance grants provide assistance for small repairs and adaptations including improvements to the thermal insulation of properties in England and Wales. Both grants are administered by local authorities. The **Home Energy Efficiency Scheme (HEES)**, which is administered by a single national agency, provides grants for loft, tank and pipe insulation, draughtproofing to doors and windows and energy advice, to eligible households in England, Scotland and Wales.

**3.2
Current
programmes**

3.2.1 HOUSING INVESTMENT PROGRAMME

In 1993/94 the total HIP for England was £1.788 billion (DoE, 1993). Equivalent figures for Scotland and Wales are £410.4 million and £78 million respectively (Scottish Office; Welsh Office). However, in England £464 million of this is in the form of Specified Capital Grants for specific programmes such as defects in non-traditional buildings, renovation of private housing, area improvements, slum clearance and improvement for sale and will not be available for energy efficiency improvements to the council stock. In addition, further sums are allocated for specific programmes including Estate Action (£356 million), Capital Partnership (£30 million) and flats over shops (£10 million). The scope of local authorities to determine their spending priorities is, therefore, limited with only £928 million of the original £1.788 billion in England being available outside central government-determined programmes. A proportion of Estates Action is spent on energy efficiency improvements and this is considered below.

3.2.2 CAPITAL RECEIPTS

Another important source of funding for local authorities is capital receipts from the sale of assets. These receipts currently stand at £5.5 billion and local authorities are allowed to spend 20% of the receipts from the sale of houses in any one year. In addition the Chancellor of the Exchequer announced in the Autumn statement in November 1992 that local authorities could spend virtually all of the receipts that accrue between the date of the statement and the end of 1993. The DoE estimates that the sales of housing during this period may generate £1750 million of receipts and if previous spending patterns are repeated this could produce approximately £1000 million in additional resources to spend on housing projects (DoE, 1993). This is an estimated figure and there is no guarantee that these sums have been realized, particularly in the current recession and the slump in house prices. One of the other features of capital receipts is that they do not necessarily occur in authorities with the worst housing conditions. Thus local authorities' ability to tackle fuel poverty, using capital receipts will be limited, unless the accumulated receipts can be redistributed and specifically earmarked for this purpose.

3.2.3 ESTATE ACTION

This scheme, which operates in England, was introduced in 1986/87 and is designed to tackle the worst housing conditions. Resources are allocated on the basis of a competitive tendering process and the programme has provided significant resources for energy efficiency improvements, although the proportion of overall funding spent on such improvements has never exceeded 25% (Table 3.2).

3.2.4 THE GREEN HOUSE PROGRAMME

This programme was introduced in 1991/92 as part of the DoE's efforts to reduce emissions of CO_2 and other greenhouse gases. The aim was to improve the energy efficiency of local authority stock and thus reduce CO_2 emissions associated with energy consumption in the properties involved. The programme was effectively ended after its second year (1992/93). Total expenditure in the two years of the programme is shown in Table 3.3.

3.2.5 PRIVATE SECTOR SUPPORT

(a) Housing Renovation Grants

These grants are provided in England and Wales under the provision of the Local Government and Housing Act 1989. There are two basic grants:

- Mandatory grants to bring properties up to a statutorily defined fitness standard.
- Discretionary grants to bring properties up to a higher discretionary standard.

Table 3.2 Estate Action Expenditure (Source: Department of the Environment)

Year	Allocation £m	Percentage spent on energy efficiency
1986/87	138	25
1987/88	191	24
1988/89	220	18
1989/90	375	23
1990/91	330	20
1991/92	373	23
1992/93	360	Not available
1993/94	390	Not available

Table 3.3 Green House Programme expenditure (Source: *Green House Programme – Final Report*)

Year	Funding	Number of local authorities	Number of schemes
1991/92	£10 million	56	74
1992/93	£45 million	65	151

In 1991/92 the total expenditure on mandatory grants in England and Wales was £249 260 000 and on discretionary grants was £22 536 000. Although there is no breakdown available of the proportion of this funding spent on energy efficiency improvements, it can be assumed that this is limited. The mandatory fitness standard, on which the majority of the expenditure is incurred, contains no reference to energy efficiency standards, therefore only a small proportion of the amount spent on discretionary grants will involve energy efficiency measures. Although Home Improvement Grants are available in Scotland, they are not targeted in any way on energy efficiency and are not mandatory for properties below the Tolerable Standard.

(b) Minor Works Assistance

The 1989 Act also provides powers for local authorities in England and Wales to provide assistance to eligible owners and tenants to carry out small-scale repair and improvement work. Eligible owners and tenants are those claiming Income Support, Family Credit, Housing Benefit and Council Tax Benefit. The work includes improvements to the thermal efficiency of dwellings and repair and replacement work to the dwelling occupied by those over 60 years of age. The latter could be used to fund improvements to, or replacement of heating systems. In the first year of the scheme only £1 610 000 was spent on grants to improve thermal efficiency in England and Wales and £6 328 000 on repairs and improvements. Once again no breakdown of this latter figure is available. These powers do not extend to local authorities in Scotland.

3.2.6 HOUSING ASSOCIATIONS

The main source of capital funding for housing associations is provided by central government via the Housing Corporation. Once again this is permission to borrow. The expenditure covers new build plus major repair work and improvements and modernization to the existing stock. The ADP for 1993/94 in England and Wales was £1.874 billion (DoE, 1993). The capital programme for associations in Scotland in 1993/94 was £261.5 million. In addition to these funds, housing associations are encouraged to raise part of the costs of their capital schemes from the private sector. In the four-year period up to 1995/96 associations are expected to raise a total of £3000 million from the private sector.

Another potential source of funding for energy efficiency improvements is via rent increases. Housing associations' assured rents on new properties are determined by the revenue needed to repay private loans and meet management and maintenance costs. Because of falling grant rates the level of private loans has increased and associations are experiencing difficulties in keeping their rents affordable. The National Federation of Housing Associations recommended that rents should be no more than 25% of net income, but the government has set the percentage at 35% contrary to the recommendation of the Select Committee on the Environment. Rents on property completed before 1988 are determined either by the rent officer or by the association itself; however, in both cases surpluses are being generated which are being used to subsidize post-1988 rents. Associations may be in a position to raise private loans to undertake energy efficiency measures which could be met by increasing rents to cover the loan repayments. Housing benefit could cover the increased rent where tenants qualify (around 60% of assured tenants do) and tenants could maintain their current heating standards at lower expenditure levels. However, rents are already too high and in some cases are dangerously close to housing benefit ceilings. Also, although increasing rent levels may not initially affect those tenants receiving full housing benefit, it will make it more difficult for them to come off benefits, thus increasing the poverty trap.

3.2.7 HOME ENERGY EFFICIENCY SCHEME (HEES)

HEES is the only current government programme specifically designed to improve energy efficiency. The legislation gives the Secretary of State at the Department of the Environment powers to make grants towards the cost of carrying out work:

- to improve the thermal efficiency of dwellings;
- to reduce or prevent wastage of energy in connection with space and water heating;
- to provide energy advice in connection with the above work.

To date these powers have been exercised to grant-aid loft, tank and pipe insulation, draughtproofing of doors and windows and energy advice. Grants are available to households in receipt of a range of benefits and those over 60 years of age irrespective of tenure.

HEES was introduced in January 1991 and, unlike other grants mentioned above, is administered not by local authorities but by a single independent agency, the **Energy Action Grants Agency (EAGA)**. Funding, and work carried out under HEES, is shown in Table 3.4.

The Chancellor of the Exchequer, in his Autumn 1993 statement,

Table 3.4 Funding and work carried out by HEES (Source: EAGA, May 1994)

Year	Total funding (no. of jobs)	Loft insulation (no. of jobs)	Draught-proofing (no. of jobs)	Combined funding (no. of jobs)	Energy advice funding (no. of jobs)
Jan-Mar 1991	£502,495 (3770)	£32,959 (195)	292,205 (2932)	£166,634 (643)	£19,696 (1063)
1991-92*	£24,256,592 (181,774)	£1,758,481 (11,050)	£13,622,258 (138,889)	£8,017,372 (31,835)	£858,490 (86,744)
1992-93*	£30,574,146 (204,711)	£1,893,679 (11,286)	£15,961,118 (150,091)	£11,489,058 (43,334)	£1,230,291 (127,987)
1993-94*	£38,499,148 (246,639)	£1,959,155 (11,178)	£20,147,867 (182,826)	£14,847,392 (55,635)	£1,551,774 (167,872)

*Figures refer to the April to March financial year.

announced as part of the scheme to compensate for the introduction of VAT on domestic fuel, that the HEES budget would be increased by £35 million for each of the next three years. In 1993/94 the total budget for HEES will be £72.5 million. This will increase the total number of homes treated by HEES to 400 000 per year. However, the eligibility for the grant was widened to include all those over 60 years of age and those claiming Disability Living Allowance. This has increased the number of eligible households from 8 million to approximately 13 million.

An evaluation of the scheme was carried out by BRECSU in 1991. The evaluation showed that HEES measures improved the NHER of a sample of dwellings examined by an average of 0.4. Average savings of £9 on annual fuel bills were also recorded in the study. The difference between this figure and the £36 average fuel bill saving predicted by the NHER audit is assumed to be benefits taken in the form of increased comfort. It is dangerous to conclude that the results were typical of HEES measures undertaken today. The work which was the subject of the study was carried out in the early days of the scheme (between May and September 1991) when the technical standards of installers and the quality of draughtproofing materials were questionable. At the time the work in the evaluation was undertaken, contractor installers in the scheme had a technical monitoring pass rate of 25% and the British Standard for draughtproofing materials was not introduced until December 1991. Despite these reservations, the evaluation appears to suggest that the loft insulation and draughtproofing materials will have a limited impact on the energy efficiency standards of those homes treated and certainly will not provide affordable warmth on their own.

3.3.1 HIP

Local authorities invest a significant level of resources in their housing stock. Their ability to make significant improvements to the energy efficiency standards is limited by the other demands on this investment, (£18.8 billion of outstanding repairs and improvements in 1985) and the proportion of spending approval allocated by the DoE to specific initiatives. Because of these two factors, unless there are significant increases in the HIP, it is unlikely that this programme will provide the level of improvement required. The DoE estimates that local authorities in England and Wales currently spend approximately £250 million of their HIP allocations on energy efficiency improvements to their stock. This figure is an estimate and no further details have been provided. Despite the fact that guidance on energy efficiency has been issued to local authorities by the Department, and priority for funding will be given to schemes with an energy efficiency element, there is no guarantee that any additional funds will be spent on energy efficiency unless the funds are distributed through specified credit approvals such as the Green House Programme.

3.3.2 THE GREEN HOUSE PROGRAMME

The Green House Programme clearly made an impact and stimulated a great deal of interest in energy efficiency amongst local authorities. The progamme was a demonstration project, which could fund models of good practice, but, with its limited resources, could not make a real impact on the problem. NEA estimates that a 15-year £17.5 billion programme is required to eliminate fuel poverty. The Green House programme provided £55 million in its two years of existence and did not operate in Scotland or Wales.

The interim results from the Green House Programme (*First Report*) indicate that in its two years the programme treated a total of 44 747 dwellings at an average cost of £1,646 per dwelling. The work produced an average reduction in CO_2 emissions of 56%. While the programme demonstrates what can be achieved, the properties treated under the programme represent a minute fraction of the 1.1 million local authority properties in England which are classified as seriously energy inefficient (unpublished BRECSU data). The DoE intended to follow up the Green House Programme with a guidance manual for local authorities and give preference to schemes which included energy efficiency measures when determining the allocation of HIP resources. Unless significant additional resources are provided, the competing demands on local authority and housing association capital resources will prevent significant increases in investment in energy efficiency measures.

3.3.3 HEES

As was noted above, there is only a limited amount of information on the benefits of HEES. The programme in 1994/95 will treat approximately

400 000 homes and the impact of the work carried out today is likely to be more effective than that examined in the evaluation because of improved standards of installation and materials. However, such a scheme which provides only basic measures at an average cost of £150, may provide valuable reductions in fuel bills and improvements in comfort but will not provide a solution to fuel poverty. The scheme has the potential to provide the solution if the Secretary of State for the Environment would exercise his powers to increase the range of grant-aided insulation measures and include replacement heating systems within the scheme. This would also require a significant increase in funding for the scheme to maintain the number of homes treated.

3.3.4 HOUSE CONDITION SURVEYS

The limited information contained in the recently published 1991 English and Scottish House Conditions Surveys make it difficult to assess the current energy efficiency standards of the nation's housing stock. Further information will be published in energy supplements to these surveys but it is unlikely that this will provide the full picture. It is clear, that from the evidence on the incidence of condensation contained in the Scottish Survey (the English Survey did not include comparable information), that a significant proportion of the stock has low thermal efficiency standards. The different bases on which the 1986 and 1991 surveys were compiled also make it difficult to assess the effect which current initiatives are having on the thermal efficiency of the stock. It is necessary to carry out a comprehensive energy efficiency profile of the nation's housing stock to assess the standard and identify those properties requiring improvement.

**3.4
Action by
the fuel
utilities and
Energy Saving
Trust (EST)**

3.4.1 FUEL UTILITIES

The privatized electricity supply companies and British Gas have a general duty to promote the efficient use of energy and provide consumers with advice on the efficient use of energy as conditions of licenses. The utilities are also required to publish Codes of Practice setting out the guidance available to customers on the efficient use of energy. In the main the utilities rely on telephone advice lines, domestic advisory staff and sales staff to fulfil their responsibilities in this area. The industry regulators do not monitor the quality of the advice provided. Irrespective of the quality of the advice, it is likely to be of limited value to low-income households who lack the capital to invest in additional insulation, new or improved heating and hot-water systems or more efficient domestic appliances. Advice alone will, therefore, not solve fuel poverty.

Under current pricing structures both British Gas and the regional electricity companies have an incentive to increase their sales of energy to maximize their profits. This so-called volume incentive will tend to discourage the utilities from investing in energy efficiency as reduced consumption will reduce their profits. This effect is somewhat negated by the competition between gas and electricity, where each utility has had to develop the efficiency of the use of their fuel to keep it competitive. Although this effect has been felt mainly in the industrial and commercial sectors, it has encouraged energy efficiency in the domestic sector. For example, the need for off-peak heating to compete with gas has resulted in the electricity companies encouraging high levels of insulation to the fabric of dwellings using the system.

It is generally accepted that the duties to provide advice contained in the utilities' licences and the incentive effect of inter-fuel competition will be insufficient to promote significant investment in energy efficiency measures. It is clear that other measures will be required. One approach would be to strengthen the duties to promote and fund energy efficiency improvements and rely upon the regulator to enforce these duties. Such an approach is likely to create its own distortions and is unlikely to be favoured by the current administration and regulators, whose approach appears to be towards less regulation of the industries.

An alternative approach would be to build in incentives to the utilities' distribution and supply price formulae which would reduce the volume incentive and enable them to increase their profits by investing in energy efficiency. The recent review of the electricity supply price formula carried out by the Director-General of the Office of Electricity Regulation (Offer) has moved some way towards this position. The volume incentive in the supply price formula has been reduced from 100% to 75% and the indications are that a similar reduction will be built into the industries' distribution price formula which is currently being reviewed.

The Director-General of Offer has also built into the supply price formula the sum of £1 per customer per year in the franchise market to fund energy efficiency measures. This will raise approximately £100m over the 4 years of the price control from April 1994. Offer has set Standards of Performance for the companies' schemes undertaken using this funding. The targets will require savings to be achieved at a cost of 1.66p per kWh or less. The Standards of Performance envisage that 35% of this funding should be spent in low-income households, although the need to provide 100% funding in this sector is likely to make the targets more difficult to achieve.

As was noted above, the electricity industry's distribution price control is currently under review. The regional electricity companies (RECs) make the majority of their profits from the distribution side of the business. The scope therefore exists in the distribution price review to increase significantly the amount of funding for energy efficiency programmes. Current indications suggest, however, that the Regulator is unlikely to provide additional funds from the distribution price control.

British Gas has a similar tariff formula to that of the electricity companies, which allows for a so-called **E factor** which would raise funds to finance energy efficiency improvements. The original concept of the E factor by the former Director-General of the Office of Gas Regulation (**Ofgas**) was that it would raise £25 million per year (to be matched by a further £25 million from British Gas profits) to finance energy efficiency improvements in the homes of low-income households. In recent months the whole validity of the E factor has been questioned by the Director-General of **Ofgas**. This situation is considered in section 3.4.2.

3.4.2 THE ENERGY SAVING TRUST

The Energy Saving Trust (**EST**) developed from an initiative proposed by the former Director-General of **Ofgas**, Sir James McKinnon. The original concept was for a Trust to channel funds from gas consumers and British Gas to finance energy efficiency improvements for low-income households. This concept was widened by the government to involve itself and the regional and Scottish electricity companies. With this expansion the Trust's proposed budget increased from the initial **Ofgas** proposal of £50 million per annum (£25 million in the form of the E factor directly from gas consumers) to £400 million per year, the majority of which was proposed to come from gas and electricity consumers.

The Trust was formally established in November 1992 and since then a number of schemes have been implemented including two of the original schemes proposed for the British Gas Trust. These are a scheme to subsidise the purchase of **gas condensing boilers** and support for small-scale **combined heat and power schemes**. The regional electricity companies also supported a Trust initiative in late 1993 which provided a subsidy to reduce the retail price of **low-energy light bulbs** by up to £5. The EEO and the Trust launched 30 **local energy advice centres** in October 1993, as a three-year pilot for a national scheme. The EEO is providing part-funding for the initiative. Although the original intention was for the LEACs to focus primarily on middle or upper-income groups and small businesses, in practice LEACs provide advice to all domestic households. None of the schemes are primarily aimed at low-income households.

A recent decision by the Director-General of **Ofgas** has called into question the whole future of the E factor and created considerable barriers for the large-scale funding of Trust schemes by gas consumers. The recently appointed Director-General has indicated that she does not believe she has the powers to raise money via the E factor and has refused to approve any additonal funds for the present schemes. The condensing boiler scheme ended in April 1994 and the combined heat and power scheme will end in June 1995. More significantly, the Trust's proposals for two major national schemes, **Energy Efficiency in Social Housing** aimed at local authority and housing associations and **Homes 2000** for owner-occupiers, have also been called into question by the decision.

Recent opinion from the Director-General of the Energy Efficiency Office suggests that this situation will require new legislation before the gas Regulator has the powers to raise funding from gas consumers. The earliest opportunity is the Gas Bill which will not be on the statute book until late 1995 at the earliest. It is therefore extremely unlikely that significant funds from the gas industry will be available for the Trust before 1996. The electricity companies can buy into Trust schemes to fulfil their obligations under the Offer Standards of Performance. Even if the whole of the £25 million per annum raised for this purpose was put through the Trust (and this is by no means likely) it will be significantly short of its projected funding.

The government is reliant upon the Trust to achieve 25% of the UK's target reductions in CO_2 emissions as part of its commitment to the Rio Earth Summit. This will create a pressure for the Trust to concentrate on middle and upper income households who have the greatest potential to reduce consumption and who are less likely to take the benefits of any energy efficiency improvements in the form of increased comfort. The Trust has clearly recognized the importance of low-income households and has specified the proportion of electricity funds to be spent in this sector. However, the importance placed on the environmental benefits of its work and the need to provide benefit to all consumers will prevent the priority being given to low-income households in the original proposals for the British Gas Trust.

3.4.3 EVALUATION

Clearly electricity and gas consumers have the ability to raise significant funds which could be spent on energy efficiency. In a full year, VAT on domestic fuel at 17.5% will raise an estimated £3 billion. Ironically, the imposition of VAT and the resultant political backlash has made it more difficult to raise significant amounts of additional money in this way. The prospects of the electricity companies or British Gas funding major programmes of energy efficiency improvements, either through the Trust or directly themselves in the short term, seem remote. Also, because of the environmental goals it is unlikely that more than 35% of any funding channelled through the Trust will be devoted to the fuel poor.

The fact that the early promise of the Trust to deliver major energy efficiency programmes look likely to remain unfulfilled in the short term is a cause of sadness and concern to all those with a general interest in energy efficiency and in the eradication of fuel poverty in particular. Unless the difficulties identified above can be resolved, sufficient funds to make an impact on the problem of fuel poverty are unlikely to come from the fuel utilities either directly or via the Energy Saving Trust.

3.5
Energy
pricing and
its effects   The price of energy will have an impact upon the level of investment in energy efficiency. The higher the price of energy, the shorter are the payback

periods for energy efficiency measures and the greater the incentive to invest in such measures. The oil crisis in the mid 1970s and the resultant increases in energy prices clearly led to increases in the efficiency with which energy was used in the commercial, industrial and transport sectors and to some extent the domestic sector. The basis of carbon or energy taxes as proposed by the EU and a number of member states is that by increasing the price of energy such fiscal measures will increase the incentive to use energy more efficiently. This argument has also been used by the UK Government as part of its justification for the imposition of VAT on domestic fuel bills.

The major difficulty with this argument in the domestic sector is that energy appears to be extremely price inelastic. That is, very large increases in price will be required to reduce the demand significantly. The further problem is that households in the low-income sector do not have the resources to make the necessary investments in the energy efficiency of their homes to reduce their consumption. It has been established that low-income households spend a higher proportion of their income on fuel than other sections of the population. People who are elderly or sick or suffer from disabilities, or are unemployed, or have young children must spend a greater proportion of their lives at home and cannot therefore rely on employers or others to subsidize their fuel costs. Also some medical conditions require sufferers to maintain adequate temperatures to prevent deterioration. Any increases in price will, therefore, adversely affect low-income households to a greater degree than the rest of the population.

It follows that any increase in energy prices must be accompanied by a programme of energy efficiency investment in the homes of low-income households. The Treasury estimates that the imposition of VAT at 17.5% on domestic fuel bills will raise £3 billion in a full year. £730 million of this will come from low-income households alone. As a minimum this latter figure should be invested in a programme to provide affordable warmth to all the 8 million low-income households in this country.

It will clearly take some time to carry out such a major investment programme. It has been estimated that a £1.25 billion per annum programme will take 16 years to deliver all the necessary improvements to the 8 million households (NEA). Such a programme must, therefore, be accompanied by temporary increases in income for fuel which will enable low-income households to pay the increased fuel bills until they are provided with affordable warmth.

3.6 European Union directives

There have been several European Union directives and initiatives aimed at improving the energy efficiency of the housing stock in member states. The two key initiatives in this field have been the proposal for compulsory **energy audits** at the time when a domestic dwelling is sold and the proposed Union-wide carbon or **energy tax**.

3.6.1 ENERGY AUDITS

The major time for investment in repair, improvement and modernization of owner-occupied housing is when the property is sold. The requirements of purchasers or funding institutions or simply the need to put the house in order so that it is more attractive to prospective purchasers, requires owners to invest in their property. The requirement to have an energy audit when a property is sold would draw attention to its energy efficiency standards. The higher the standards, the more marketable the property would become. Prospective owners would start to consider the running costs of properties and the most energy efficient properties could command higher prices, as the higher mortgage repayments would be offset by lower fuel bills. Owners would be encouraged to improve the energy efficiency standards of their properties to improve their marketability.

The move to introduce mandatory energy audits at the point of sale has been resisted by the British Government on the grounds that it would represent an unacceptable barrier to the sale and purchase of houses. Unless there is a complete change of policy in this area, it is unlikely that such a mandatory requirement will be introduced in Britain. Even if it were it is unlikely to have a significant impact on fuel poverty in the short- and medium terms. By encouraging the increase in general standards of energy efficiency in the housing stock, all sections of the population will benefit. However the most immediate benefits will accrue to owner-occupiers with the necessary resources to invest in their property. It was noted above that low-income owner-occupiers do not have the necessary capital to improve the energy efficiency of their houses. In the short-term such a move is likely to trap low-income owner-occupiers in energy inefficient housing by making such houses difficult to sell.

3.6.2 ENERGY TAX

The proposals for a Union-wide carbon or energy tax would make energy efficiency investment more attractive and reduce payback periods. A recent decision by the Greek presidency of the Union to drop the proposal for such a tax means that it is unlikely to be introduced in the foreseeable future. Such a move on its own would be unlikely to improve the situation of low-income households in fuel poverty. Once again such households do not have the necessary capital to reduce their energy consumption. If, however, a significant proportion of the revenues from such a tax was invested in energy efficiency improvements for low-income households and transitory relief were given to compensate for the increased energy prices, such an initiative could make a significant impact on fuel poverty.

The recent decision by the British Government to introduce VAT on domestic fuel has highlighted the sensitivity of using fiscal measures to increase fuel prices. It has also demonstrated the potential for such measures to raise revenue. As was noted in section 3.5, when VAT is levied at 17.5% from April 1995, it is estimated that £3 billion will be raised annually. This is clearly sufficient to finance a major energy efficiency programme for low-income households and to provide increased income support before an energy efficient home can be provided to all low-income households.

3.6.3 EU FUNDING SOURCES

The European Union also provides a number of funding sources for energy efficiency demonstration projects. The current programmes are:

- **SAVE**, a programme that aims to examine institutional areas which if amended could lead to improved energy efficiency.
- **Thermie**, which aims to demonstrate novel energy efficiency technologies and to desseminate the results of successful demonstration projects and assist technologies that have been demonstrated as effective, to gain greater market acceptance.
- **Joule**, which supports basic research and development in both hardware and software.

These programmes offer average funding of 40% of project costs. While they offer a potential for part of the funding for demonstration projects and are undoubtedly a valuable source of information on techniques and methods, they will not provide the level of funding necessary to solve the problem of fuel poverty.

In addition, from 1st January 1995, the European Union will introduce a mandatory system of energy labelling for domestic refrigerators and freezers. The importance of appliance efficiency is considered in Chapter 4. While this will have important implications, because of the difficulty which low-income households have in purchasing new domestic appliances, it is unlikely to impact upon this section of the community in the short-term.

**3.7
Energy audits**

Energy audits provide an objective means of measuring the energy efficiency of properties and setting targets for improvements. There are two main energy rating services currently in operation in Britain:

- **National Home Energy Rating Scheme (NHER)**, administered by the National Energy Foundation and providing a 0-to-10 rating to one decimal point (0 being the least efficient and 10 the most).
- **Starpoint**, administered by M.V.M. Starpoint, which provides a five-star rating (one star being poor and five stars being the most energy efficient).

Both schemes provide a variety of other information including CO_2 emissions, fuel costs and options for improvement.

In July 1993 the government introduced its **Standard Assessment**

Procedure (SAP) which provides a means of estimating the energy efficiency performance of dwellings on a scale of 1 to 100 (1 being poor and 100 being the most energy efficient). The SAP also provides a method of comparing the ratings provided by the two schemes outlined above. Both the NHER and Starpoint now include a SAP rating in their results. Other rating schemes are currently under development. It is important to potential users to examine the available systems and decide which of the additional features provided best suit their requirements. In terms of affordable warmth, it is important to consider the actual costs of maintaining adequate temperatures within the home. For this reason, rating systems which take account of the different climatic conditions throughout the UK are preferred.

Ratings can be provided on single dwellings, both existing and proposed. In addition, large numbers of dwellings can be rated using stock profile versions of the audits which are particularly useful for landlords with a large number of properties. Although these stock profiles provide less accurate results than full audits on individual properties, they can still provide valuable comparable information across a large number of properties.

Energy audits are not initiatives which will reduce the incidence of fuel poverty in themselves. They are, however, a valuable tool to assess the energy efficiency of dwellings, prioritize the worst property, examine the options for improvement and provide a measure of the rate of improvement. Additionally, if the ratings provide fuel cost details, they enable affordable warmth targets to be set.

3.8 Building Regulations and other standards

3.8.1 BUILDING REGULATIONS

The Building Regulations control the standards in newly constructed or significantly adapted dwellings. Since 1965, the Building Regulations have included thermal efficiency standards. Clearly the effect of such standards on the housing stock will be limited given the current rate of replacement of the existing stock. Some 67.1% of the present housing stock was built before 1965 and, given current demolition rates of approximately 3358 per year and new building at 167 724 units per year, some 54% of the housing stock in 30 years time will have been built prior to 1965 (*Housing Construction Statistics*, 1991, 1992; *English and Scottish House Condition Surveys*, 1991).

The majority of new house building is in the owner-occupier sector and as such will not be generally available to low-income households. The only newly built houses available to them will be a proportion of the 33000 housing association properties built for rent (Housing Corporation, 1993–94). Local authority new building for general housing need has virtually ceased.

The Building Regulations do not apply to major refurbishment of existing dwellings and would only apply to sections of existing buildings which were rebuilt or newly built. Thus an existing house which is refurbished would be unaffected by Building Regulations but a newly built extension would be subject to the conditions of the regulations. An existing home could be refurbished without any requirement to bring the energy efficiency standards up to those of current Building Regulations.

A house built to current Building Regulations will score approximately 7 on the 0-to-10 scale of the National Home Energy Rating Scheme, while the national average of the current stock is between 4 and 5. Unless the Building Regulation Standards can be extended to the existing stock, they will have a limited influence on energy efficiency standards in the stock as a whole. Without a significant intervention in the existing stock, such as a major repair or improvements scheme, it is difficult to see how Building Regulations could be extended to the existing stock. In each revision of the Regulations since 1965 the energy efficiency standards have been increased. The current proposed revision to the 1990 Regulations, in addition to recommending increased standards of individual components, also for the first time includes the efficiency of the heating system used. The proposals also suggest the extension of the Regulations to major alterations and conversions.

Despite the improvements in their intent and scope, by their very nature (as noted above) Building Regulations will have only a minor impact on the majority of the housing stock. It should also be noted that even the proposed increases in standards are far below those in other Northern European countries and these proposals have encountered significant opposition from the private house building sector which sees them as an unwarranted intrusion into the freedom of private house builders and purchasers. The Regulations clearly have a long way to go before they have any significant impact upon fuel poverty.

The Building Regulations have another important rôle in that they set standards which are used when carrying out work which does not come under the Regulations. Standards for grant-aided work and work carried out by individuals and public bodies are often tied to Building Regulation Standards. The Regulations act as a model of good practice for those carrying out building work.

3.8.2 THE FITNESS FOR HUMAN HABITATION STANDARD (ENGLAND AND WALES) AND THE TOLERABLE STANDARD (SCOTLAND)

Central to the concept of the rehabilitation of existing dwellings has been the concept of a basic standard to which properties should be improved. These standards are the Fitness for Human Habitation Standard (which applies in England and Wales) and the Tolerable Standard (which applies in Scotland). Failure to meet these standards opens a landlord to legal action by the local authority both to bring the property up to the standard and to provide the tenant with compensation. These standards are also the trigger for access to mandatory improvement grants as noted in section 3.2.5(a).

Both standards make reference to the provision of facilities to heat the dwelling and also to freedom from dampness. In the case of the Fitness for Human Habitation Standard (Current Legislation Local Government and Housing Act 1989) these are that:

- it is free from dampness prejudicial to the health of the occupants (if any);
- it has adequate provision for lighting, heating and ventilation.

The Tolerable Standard (defined in the Housing (Scotland) Act 1987) states that dwellings must amongst other requirements:

- be substantially free from rising or penetrating damp;
- have satisfactory provision for natural and artificial lighting, for ventilation and for heating.

The standard has not changed in respect of these two points since its introduction in the Housing (Scotland) Act 1969.

Under the system in England and Wales, the definition of dampness has been extended beyond rising and penetrating damp, to include dampness prejudicial to the health of the occupant – a definition which clearly includes dampness resulting from condensation. Indeed the guidance on the 1989 Act provided by the Department of the Environment, in *Circular 6/90*, clearly identifies dampness resulting from condensation as coming within the scope of the Act. Since the principal causes of condensation are the lack of adequate heating (or the inability of the householder to afford to heat the home to adequate temperatures) and thermally inefficient dwellings, this definition clearly has the potential to have a significant impact on the problem of fuel poverty. No such widening of the definition of dampness is included in the guidance which accompanies the Tolerable Standard.

The legislation in Scotland and the guidance provided to local authorities to enforce it will clearly not tackle fuel poverty. It remains to be seen whether the legislation in England and Wales is enforced in the same way suggested in the guidance. It should be remembered that enforcing such standards implies the provision of mandatory grant aid. Many local authorities may, therefore, be reluctant to extend the definition of unfitness to include condensation when they are already concerned about the proportion of their grant aid budget which goes to mandatory grants.

In any event, such legislation can be applied only to the private rented sector which, although it represents approximately 1.5 million households, still represents only a small proportion (10.3%) of the households suffering from fuel poverty.

3.9 Summary

It is clear that the current level of expenditure on the nation's housing stock will not solve the problem of fuel poverty. There is evidence to suggest that the current expenditure programmes are not keeping pace with the non-energy efficiency demands made upon them. Demonstration projects, such as the Green House Programme and Estates Action, provide useful examples of what can be done; however, unless they can be applied across the housing stock as a whole the impact of such demonstration schemes will be limited.

The accumulated capital receipts held by local authorities and sums raised by the imposition of VAT on domestic fuel demonstrate that the additional resources required to eradicate fuel poverty can be found, if there is the will to solve the problem. Apart from central government, the only other potential funding sources which could raise the level of funding required would be the fuel utilities. Current indications are, however, that a change in legislation will be required before the utilities could raise the level of funding required. This seems to be a step the government is unwilling to take at the present time.

References

Developing Local Authority Housing Strategies – Audit Commission, 1985.
Department of the Environment *Annual Report* 1993.
Green House Programme – Final Report, 1993.
Interim Evaluation of the Home Energy Efficiency Scheme – BRECSU, 1993.
Warm Homes, Cool Planet – A joint NEA, FoE, Heatwise Glasgow and National Right to Fuel Campaign Report.
English and Scottish House Conditions Surveys 1991.
Housing Construction Statistics 1991.
Housing Construction Statistics 1992.
The Approved Development Programme 1993-94 – The Housing Corporation.

Appliances and affordable warmth 4

Marcus Newborough

In the domestic sector, each of us requires thermal comfort and what might perhaps be termed 'other comfort' facilities. **Thermal comfort** is a strong function of the thermal design of the building and its space-heating system, but the **other comfort facilities** depend mainly upon the availability of suitable domestic appliances within the building. These 'appliances' range from toasters to televisions and from cookers to coffee makers. In the context of the fuel poor, it is desirable to raise thermal comfort levels by improving the energy performances of buildings, space-heating systems and (indirectly) appliances. By reducing energy expenditures on appliances, the savings achieved may be redirected to buying more warmth and/or energy-saving building improvements. The means for achieving, and significance of, this process are reviewed here.

It should be noted that, in this chapter, energy cost predictions are based on unit prices of £0.02 and £0.08 per kWh for natural gas and electricity respectively, and all data exclude the 'standing charge' element of a household's energy bill. It is assumed that average use is made of average appliances in average climatic conditions and that off-peak tariffs are not utilized to reduce electricity bills. Where a particular saving has been estimated on the basis of switching to an off-peak electricity tariff, a unit price of £0.03 per kWh has been asssumed.

**4.1
Summary**

**4.2
The appliance-
purchase/energy-
use interface**

When compared with the rate of domestic new builds and refurbishments, the energy-consuming devices employed in dwellings are replaced relatively frequently in order to meet real or perceived needs. Existing appliances are, on average, replaced every 6 to 12 years depending upon factors such as appliance type, usage and disposable income. An average household invests in a new appliance (i.e. a brand new or secondhand one) as a replacement device every 1–2 years. Also, we occasionally purchase appliances (such as tumble driers, microwave ovens and electric blankets) which previously have not been fitted in the dwelling and sometimes buy an additional model (e.g.

a second TV). This high turnover of the appliance stock provides an opportunity for consumers to reduce domestic energy bills by obtaining more efficient equipment at the points of purchase. Additionally, the household may realize savings by implementing 'good housekeeping' practices with respect to appliances. To achieve these savings, more efficient appliances have to be readily available in the UK market-place together with facilities for encouraging individuals to make discerning purchasing decisions partly on the basis of energy performance data (e.g. appropriate labelling, leaflets, media advertising and advice programmes).

There are two main energy consumption trends with respect to the appliance market. The first trend is towards producing appliances which are more energy efficient. In general, the rate of efficiency improvement has been slow due to the lack of incentives for manufacturers to produce high efficiency versions of their products. This is because the appliance market is largely unregulated in the UK and unit energy prices are low in real terms. Ironically, in a few cases, recent design changes have increased energy consumptions (e.g. the power inputs of vacuum cleaners have been increased, and lower external surface temperatures have been achieved for some electric ovens by forcibly cooling the oven's thermal insulation). However, EU-backed ecolabelling and energy labelling for appliances is a significant driving force; more efficient equipment (in all product categories) will eventually become available.

The second trend is the well-established one of manufacturers introducing new types of 'desirable' white and brown goods, marketed in some cases with the support of the electricity and gas industries (e.g. dishwashers and gas-fired tumble-driers). To survive commercially in what is becoming an increasingly saturated market (Table 4.1), appliance manufacturing companies need to innovate and market new products. Similarly, the domestic consumer market for electricity and gas is very important to the competing energy-supply industries. With some products (e.g. compact fluorescent lights and Economy Seven water heaters) total expenditures will be reduced if the new appliance replaces its traditional counterpart. Conversely, a household making a first purchase of a tumble drier or dishwasher will augment significantly its annual electricity bill. In other cases, the probability of expenditures increasing or decreasing is less clear, due to the way in which the household may employ the new appliance. For example, automatic electric kettles with water-height scales offer scope for saving energy, but the automatic feature reduces the urgency, otherwise associated with conventional kettles, of pouring the water as soon as it has boiled. Consequently, there tend to be more electricity-use events per unit volume of boiled water employed.

Overall, it is clear that UK households purchase several million appliances each year and that new innovations tend to augment, rather than reduce, domestic energy bills due to their labour-saving and/or additional (rather than replacement) rôle in the home. It is therefore important to provide

Table 4.1 Percentage of homes owning stated electrical appliance for average, economically inactive and professional households (*Social Trends*, 1994, and GK Marketing Services Ltd., 1994, unless stated otherwise)

	Average	Economically inactive	Professional
Washing machine	88	79	96
Tumble drier	49	32	66
Microwave oven	59	41	68
Dishwasher	16	7	41
Television	99	99	99
Freezer and/or fridge-freezer	85	76	94
Fridge	51	[48]	[54]
Freezer	38	[36]	[40]
Fridge-freezer	54	[41]	[52]
Lights	100	[100]	[100]
Iron	98[a]	[98]	[98]
Vacuum cleaner	97	[97]	[97]
Kettle	86[a]	[86]	[86]
Cooker	41[b]	[41]	[41]
Water heater - 12 months p.a	{32}	[32]	[32]
Water heater - summer-only users	{13}	[13]	[13]
Instantaneous space heater(s)	{5.5}	{8}	{2}
Miscellaneous	—	—	—

Notes
In general the stated ownership levels are assumed to imply that active use is made of owned appliance by the household.
Numbers in [ ] indicate assumption based on data for average household and any relevant available information for distribution with socioeconomic group.
Numbers in { } indicate estimate based on data from Evans and Herring, 1990.
a From Henderson and Shorrock, 1990.
b Approximately 30%, 15% and 9% of households own free-standing electric cookers, built-in ovens and built-in hobs respectively. This is interpreted as 41% of households having free-standing electric cooker equivalents (ownership distributions with individual cooking-appliance type not available).
— Ownership data with socioeconomic group exist for some low-energy consuming appliances such as VCRs, home computers and CD players, but not for more significant consumers (e.g. toasters, showers and second/third TVs).

individuals with the options of:

• obtaining high-efficiency appliances, irrespective of whether they are replacing within, or adding to, their household's stock of appliances; and
• using existing or new appliances in a manner that will lead to lower annual energy bills without having a regressive effect upon lifestyles.

**4.3
Ownership
and utilization**

The distribution of appliances in a home, the degree to which they are utilized and the preponderance of electrical equipment vary significantly with socioeconomic group (Table 4.1). In professional households dishwashers are common (41%), but in economically inactive ones they are rare (7%). Conversely, poor households living in non-centrally heated (NCH) homes rely upon 'instantaneous' room heaters (e.g. electric fan heaters and paraffin heaters) more than professionals whose homes tend to be centrally heated. For example, 30% of DEs and 7% of ABs who live in NCH homes depend upon electrical room heaters (Evans and Herring, 1990). In some homes, especially those of the elderly poor, appliance utilization levels may be well below average because of the existence of a thrift ethic involving, for instance, precise use of kettles, lights, televisions and saucepan lids. Equally research shows that, when attempting to be thrifty, individuals often pay too much attention to actions which will have little beneficial effect upon their annual energy bills (such as obsessively switching off lights and closing refrigerator doors quickly). This is because of the lack of understanding of the **relative** energy consumptions of appliances (Stern and Aronson, 1984; Hedges, 1991). Indeed most of us are unsure about the financial (dis)benefits of, say, keeping freezers well filled, using microwave ovens in place of hobs and using kettles in place of immersion heaters. It may be argued that it is in the vested interests of the appliance manufacturers and gas and electricity industries for this type of uncertainty to continue, but the aim here is to shed some light.

Most domestic appliances are electrically driven; the application of gas-fired equipment in dwellings is limited mainly to boilers and cookers. The end-use efficiencies of electrical appliances tend to be significantly higher than their gas-fired equivalents, but the efficiency advantages are rarely sufficient to counteract the financial disbenefits associated with utilizing high unit-cost electricity. Therefore, if significant financial savings are to accrue to consumers, efforts should be focused mainly on reducing the consumptions of major electrical appliances (the definition of 'major' being an appliance with a typical annual consumption of at least 100 kWh).

On average, electricity use per domestic customer increased by 250 kWh per year in the UK between financial years 1981/82 and 1991/92 (Sterlini, 1993). At present, the total electricity consumption of the UK domestic sector is ~100 TWh, with predicted annual electricity expenditures per average household of ~4.39 MWh (£351) including ~2.66 MWh (£213) for running appliances (see later).

**4.4
Energy
consumptions
and savings**

To define accurately the energy spent on appliances among poor households, relative to the national average, a comprehensive monitoring programme to identify (by socioeconomic group) the typical annual electricity consumptions of each appliance type is desirable. However, the cost of such a survey is prohibitive and so estimates have to be made on the basis of consumption data available from various sources. In some cases this information may be a few years old, unrepresentative of the average household or not entirely free from vested interests.

Average annual electricity consumptions and approximate running costs per appliance type are compiled in Table 4.2. From this data and the available appliance ownership information in Table 4.1, average annual electricity costs per household type have been determined (Table 4.3). The variations in average annual spend between household types need to be treated with caution, due to the averaging process and the accuracy of the ownership and consumption data (Tables 4.1 and 4.2). A better approach is to predict the probability of a household possessing a certain stock of appliances from the ownership data and then estimate average annual electricity consumptions for various stocks (Table 4.4).

By defining feasible means for reducing the costs of running appliances (Table 4.5), achievable savings can be predicted for household types having various stocks of average appliances (Table 4.6). Of course making estimates of what can be saved assumes that the household has not already adopted the energy efficient practice in question – if it has then the total achievable savings will be reduced accordingly.

4.5 Discussion

Energy expenditures on comfort facilities (such as the provision of hot water, cooked food, washed/dried clothes/dishes, hot drinks and TV) are largely independent of the thermal design of the occupied dwellings. Much of the energy consumed by appliances is dissipated as wild heat within the indoor environment, but this incidental space-heating gain is of dubious value. Besides being expensive when obtained via electrical appliances, such heat inputs are unwelcome in mild weather. In kitchens, the rate of heat gain from cooking equipment is high and tends to occur during peak activity periods when it is least required. Also, gains from appliances such as washing machines, tumble driers and cookers are accompanied by unwanted increases in indoor humidity levels. Therefore, there is little merit in the argument that waste heat emanating from appliances is beneficial; preferably each appliance should be designed to achieve its primary purpose efficiently (and hence cheaply) and the end users should achieve adequate levels of thermal comfort by virtue of their occupying thermally-efficient buildings, which are equipped with effective space-heating systems.

In general there is a wide distribution in annual energy consumptions per household due to differences in the benchmark efficiencies of appliances designed to achieve the same task, and to variations in appliance-use behaviour among consumers. Some estimate of the variations in energy performances can be obtained by referring to standard laboratory test results reported in the literature (i.e. behaviour-independent data). Firstly, an EEO-approved list of 57 refrigeration appliances by product type provides a 1-to-10 rating of 'bad' to 'good' for labelling the appliances with respect to their energy requirements (Eastern Electricity, 1993). Upon analyzing the data, a wide

Table 4.2 Average annual electricity consumptions by appliance type (derived from data provided by Evans and Herring, 1990; Shorrock *et al*; 1992; *Social Trends*, 1994; EEO, 1994; Electricity Association, 1994)

Appliance	Estimated average annual expenditure	
	(kWh)	*(£)*
Water heater	2736; 855[a]	219; 68
Cooker	846	68
Freezer	732	59
Fridge-freezer	615	49
Dishwasher	482	39
Lights	360	29
Fridge	357	29
Kettle	259	21
Tumble drier	254	20
Television	209	17
Washing machine	205	16
Iron	78	6
Microwave oven	75	6
Vacuum cleaner	26	2
Miscellaneous	200[b]	12
Space heating	700; 4000; 5300[c]	56; 320; 424/159

Notes

Where feasible, averages have been calculated based on available data to form the kWh per annum estimates. The least reliable estimates are those which have been derived from national aggregate data (see notes a, b and c below). However, in all cases a wide tolerance band (e.g. +/− 50%) should be expected, because appliance energy performances within a certain product category usually vary substantially depending upon the end user, the manufacturer and in some cases the age of the appliance. All cost estimates assume that cheap off-peak tariffs are not utilized (but see note c below).

a   First figure is the average for homes relying on an electric water heater 12 months p.a.; second figure is the average for those homes switching to electric water heating only during the summer.

b   Aggregate national consumption shared equally by all households.

c   First figure is the average for non-centrally heated (NCH) homes employing electric room heaters as supplementary heaters; second figure is the average for NCH homes relying totally on said heating means; third figure is average for electric centrally heated (CH) homes (74% of which employ storage heaters). From Evans and Herring (1990) and Shorrock *et al*; (1992), about 7.3 million homes rely on electricity for space heating, of which 2.0 million are centrally heated. Of the NCH homes, 0.6 million rely solely upon electric room heaters, leaving 4.7 million who use electricity as supplementary heating means. Therefore, 4.7 million NCH households use these appliances to boost their indoor comfort levels above those achieved by the main form of space heating (e.g. coal fires). On average, CH and NCH homes achieve respective mean indoor air temperatures some 9.8 and 7.3 °C higher than mean external temperatures during winter (1989 data from Shorrock *et al*; 1992). For lack of other information, the analysis here makes an assumption that, on average, the 4.7 million supplementary users realize an additional 1.25 °C rise from electric room heaters above the average ΔT of 7.3 °C achieved by NCH homes. Therefore, about 65% of national electricity consumption for space heating is dedicated to CH homes, 15% to NCH homes where electrical room heaters are the only form of heating and 20% to NCH homes where such heaters play a supplementary role. Estimates of total annual expenditure in Great Britain on electricity for domestic space heating range from 12.6 TWh to 20.3 TWh. Taking an average of 16.45 TWh, it is predicted that NCH homes relying solely on

Table 4.3 Predicted average annual electricity expenditures for average, economically inactive and professional households.

Household type	Expenditure per annum (kWh)	
	On appliances	Total
Average	£213 (2660)	£351 (4390)
Economically inactive	£195 (2430)	£345 (4320)
Professional	£229 (2860)	£356 (4450)

Notes

The expenditure on appliances does not include water heaters of instantaneous space heaters. Electricity expenditures are influenced considerably by: (i) the extent to which electric water and/or space heaters are employed; (ii) the exact stock of appliances at the disposal of the household (see Table 4.4); (iii) the way, and the frequency with which each appliance is used; and (iv) how informed and thrifty the end-users are with respect to energy use.

Table 4.2 Notes continued

electric room heaters spend ~4000 kWh p.a. and supplementary users in NCH homes spend 700 kWh p.a., while homes with electric CH spend ~5300 kWh p.a. Two cost estimates are stated for the latter assuming that all the electricity for space heating is bought via normal or off-peak tariffs.

It is implicit in this analysis that NCH homes which rely upon electric room heaters require (on average) an additional 34% of space-heating in order to achieve the same mean indoor air temperature as an average CH dwelling. Affording this increased level of warmth (without changing the type of space heating) will, on average, cost 1360 kWh (£109) per annum, assuming that the conversion of electricity to heat is 100% effective and zero difference between the thermal design of an average CH and NCH dwelling; if not, this additional heat requirement could be substantially higher.

Table 4.4   Probability of electrical appliances being at the disposal of stated household type and the associated annual energy costs

Household code	Appliance stock	Percentage probability of owning at least the stated stock of appliances if household type is:			Annual cost of running stock (£)
		Average	Economically inactive	Professional	
A	TV, L, I, VC	94	94	94	54
B	TV, L, I, VC, K	81	81	81	75
C	TV, L, I, VC, K, WM	71	64	78	91
D	TV, L, I, VC, K, WM, MO	42	26	53	97
E	TV, L, I, VC, K, WM, FF	38	26	40	140
F	TV, L, I, VC, K, WM, F	36	31	42	120
G	TV, L, I, VC, K, WM, F, FSC	15	13	17	187
H	TV, L, I, VC, K, WM, F, FSC, FR	6	5	7	246
I	TV, L, I, VC, K, WM, F, FSC, FR, MO	3	2	5	252
J	TV, L, I, VC, K, WM, F, FSC, FR, MO, TD	2	0.6	3	272
K	TV, L, I, VC, K, WM, F, FSC, FR, MO, TD, D	0.3	0.04	1	311
L	TV, L, I, VC, K, WM, F, FSC, FR, MO, TD, D, FF	0.2	0.02	0.5	360

Key
D = dishwasher; F = fridge; FF = fridge-freezer; FR = freezer; FSC = free-standing cooker (or equivalent); I = iron; K = kettle; L = lights; MO = microwave oven; TD = tumble drier; TV = television; VC = vacuum cleaner; WM = washing machine.

Table 4.5 Options available now for reducing the annual cost of running
the stated appliance.

Appliance	Action	Action code	Estimated annual saving
Lights	Convert to fluorescent	1	270 kWh (£22)
Washing machine	Increase use of low temperature wash cycles and shift 25% of use to off-peak tariff	2	50 kWh (£4)[a]
Fridge*	Change to an off-peak tariff	3	£5[a]
	Replace with most efficient model	4	65 kWh (£5)[b]
	Replace with most efficient larder fridge	5	120 kWh (£10)[c]
Freezer*	Change to an off-peak tariff	6	£11[a]
	Replace with most efficient upright model	7	367 kWh (£29)[b]
	Replace with most efficient chest freezer	8	480 kWh (£38)[b]
Fridge-freezer*	Change to an off-peak tariff	9	£9[a]
	Replace with most efficient model	10	155 kWh (£12)[b]
Free standing cooker	Replace with gas cooker	11	£45[d,e]
Tumble drier	Switch 50% of utilization to off-peak tariff	12	£7[a]
	Replace with gas tumble drier	13	£13[e]
Dishwasher	Switch 50% of utilization to off-peak tariff	14	£10[a]
Kettle	Use gas kettle	15	£10[e]
Water heater	Switch 33% of utilization to off-peak tariff	16	£45[a]

Notes
a Estimate. Any change to an off-peak tariff has to be considered with respect to the associated changes to the standing charge and day-time unit price. If the day-time unit prices for the off-peak and normal tariffs are identical, the household needs to spend £10–£20 of the savings on the extra standing charge. If not, a calculation is required based on the household's annual electricity consumption, the available tariff(s) and the number of 'actions' being considered for implementation.
b Straightforward replacement of existing appliance with model which has the least annual electricity consumption (see *).
c Replacement of existing fridge with most efficient version which does not have an ice box (assumed in Table 4.6 to be relevant only for households already owning freezers and/or fridge-freezers).
d The following assumptions have been made to obtain an average saving:
   (i) Average thermal efficiencies: gas hob = 44%; gas oven = 8%; electric hob = 58%; electric oven = 13%. (Subdivision of the data to identify that used for grilling operations is not feasible.)
   (ii) The average annual electricity consumption for a free-standing cooker of 846 kWh is split 677 kWh for the hob and 169 kWh for the oven. This equates to respective thermal-energy requirements of 393 kWh and 22 kWh, which require 892 kWh and 275 kWh of gas if achieved

variation in annual energy consumptions is apparent. For example, if consumers use the list to buy the most efficient fridge they will pay £0.18 p.a. for each (gross) litre of its capacity, while the worst model will cost £0.33 per litre per year. Selection of the best rather than the worst listed model of upright freezer will cost £0.18 as against £0.67 per litre per annum. Although these are the extreme range values, the 1–10 is not necessarily effective at delineating running cost variations between similar appliances with similar ratings. For example, the respective annual costs of running three of the listed larder fridges, each with a rating of 8, amount to about £0.14, £0.12 and £0.09 p.a. per litre of capacity. If a consumer chooses the only appliance with a rating of 9, this annual cost decreases to £0.07 per litre (i.e. about half the cost of running the worst appliance with a rating of 8). A rating on an energy label should not require prospective purchasers to divide annual energy requirement (in kWh) by capacity (in litres) in order to make sound 'value for money' comparisons. In addition, it appears that the effectiveness of the current scheme is being compromised by its voluntary, rather than mandatory, nature (e.g. only the retail outlets of the regional electricity companies are participating). At present, these energy labels tend to receive a low priority in the purchasing decisions made by consumers.

Of the appliances which are currently 'unlabelled', some useful running-cost data is available for groups of ovens and washing appliances (Consumers Association, 1991). For ovens, the prescribed test was to determine the energy requirement for maintaining the (empty) oven at 200 °C for a period of one hour immediately following the pre-heat period. The results were: gas ovens, 0.023p to 0.055p per litre; electric ovens, 0.073p to 0.154p per litre. For comparison, a prototype high-efficiency domestic electric oven recently developed at Cranfield was subjected to the same test; it consumed ~0.03p per litre (Newborough, 1994). The following ranges of annual running costs were also reported: full-size dishwashers, £42 to £62; full-size front-loading washing machines, £12 to £31; tumble driers, £19 to £29; and washer-driers, £33 to £60. An investigation of toasters (Newborough et al; 1987) revealed that electricity consumptions ranged from 0.9 to 2.8p per operation for 2 slices of unfrozen bread. Evidently, if consumers are to become aware of their options with respect to buying more efficient equipment, it is important to apply legitimate energy labels to each type of major appliance.

The average user can usually achieve significant savings by modifying appliance utilization practices. The effectiveness of this depends upon the

Table 4.5 Notes continued

    with a gas hob and oven. Thus the predicted annual cost of electric and gas cooking is £68 (846 kWh) and £23 (1167 kWh) respectively, suggesting a typical saving of £45 p.a. If the household only switches to a gas hob, the predicted annual saving is £36.

e   Combustion products from gas-fired appliances, such as hobs, enter the indoor environment directly: there is some evidence that this may have a very small effect on the human respiratory system (Ogston et al; 1985; WHO, 1990).

*   Based on Eastern Electricity data for the EEO's energy efficiency rating of refrigeration appliances.

Table 4.6 Options available for reducing annual electricity costs by household code (refer to Tables 4.4 and 4.5)

Household code	Action codes implemented	Predicted financial saving achievable per annum (£)	Existing annual expenditure neglecting miscellaneous uses* (£)	Proportion saved(%)
A	1	22	54	41
B	1,15[a]	32	75	43
C	1,2,15[a]	36	91	40
D	1,2,15[a]	36	97	37
E	1,2,9,15[a]	45	140	32
	1,2,10,15[a]	48		34
F	1,2,3,15[a]	41	120	34
	1,2,4,15[a]	41		34
G	1,2,3/4,11,15	86	187	46
H	1,2,3/4,6,11,15 (min)	97	246	39
	1,2,5,8,11,15 (max)	129		52
I	1,2,3/4,6,11,15 (min)	97	252	38
	1,2,5,8,11,15 (max)	129		51
J	1,2,3/4,6,11,12,15 (min)	104	272	38
	1,2,5,8,11,13,15 (max)	142		52
K	1,2,3/4,6,11,12,14,15 (min)	114	311	37
	1,2,5,8,11,13,14,15 (max)	152		49
L	1,2,3/4,6,9,11,12,14,15 (min)	123	360	34
	1,2,5,8,10,11,13,14,15 (max)	162		45

Additional annual saving if household relies entirely upon immersion heater (£)	45

Notes

* Data is not available to define miscellaneous use for individual household types: it is predicted to be £16 for an average household. However, the miscellaneous category may account for an additional £5 to £50 p.a. depending on the utilization of small appliances (such as toasters and hair driers), and second appliances (e.g. televisions and stereos). Similarly, achievable financial savings in the miscellaneous category may range from negligible to say £12 p.a.

a assuming a gas (rather than a solid-fuel or no) cooker is fitted in the dwelling.

appliance type, the quality of information/advice available and how receptive each member of the household is to saving energy/money. A summary of general energy efficiency advice with respect to appliances is provided in Table 4.7. Certain changes are highly desirable and actions should be prioritized in descending order of annual energy consumptions (Table 4.2). For example, fitting a 50 mm thick segmented insulation jacket to a bare 210 litre immersion tank will lower its standing losses from ~23 kWh per day to ~5 kWh per day (Stephen, 1986). Similarly, a new factory-insulated tank with a 50 mm cladding will yield ~2.7 kWh per day while a 150 mm layer of insulant will lower the standing losses to ~1 kWh per day. Other behaviour changes are desirable, e.g. turning down the energy supply to a hob as soon as the water has boiled and ensuring that a lid is fitted to the pan whenever feasible (Newborough and Probert, 1988). Also, manually defrosting a freezer (or the ice box of a conventional fridge) at regular intervals is worthwhile. Switching to an off-peak electricity tariff and attempting to utilize, say, the washing machine, tumble drier and/or dishwasher (partly) during the associated cheap period is advantageous, even if timers and smoke alarms need to be purchased (Deering *et al*, 1993). However, it should be remembered that even the most energy-conscious user is limited by the inherent efficiency of the available equipment when practising self-help to modify existing usage practices in line with energy efficiency advice.

4.6 Conclusions

This investigation has concentrated on predicting annual expenditures and achievable savings with respect to domestic electrical appliances (excluding any savings that might be achieved with respect to electric space-heating appliances and miscellaneous equipment). The predicted average expenditures on appliances by socioeconomic group (Table 4.3) are based on the associated appliance ownership data (Table 4.1) and available energy consumption data which assumes average utilization of average appliances (Table 4.2). Expenditures are also predicted for particular stocks of appliances within a home (Table 4.4), and the associated estimates of achievable savings (Table 4.6) are based on 'actions' available to consumers for reducing their annual costs of running appliances (Table 4.5). In all cases, the predicted savings may be augmented by implementing good housekeeping measures (Table 4.7).

If the use of electrical space and water heaters is neglected, the average annual spend on appliances is estimated to range from 2430 kWh (£195) for an economically inactive household to 2860 kWh (£229) for a professional household. An otherwise average household (Table 4.3) occupying a non-centrally heated dwelling, which makes average use of electric water heating 12 months p.a. and of electric instantaneous room-heating appliances, is predicted to spend ~9400 kWh (£752) annually.

Predicted savings in appliance energy bills for various stocks of appliances (Table 4.6) range from £22 p.a. for a very poorly equipped home to £162

Table 4.7 Rule of thumb 'good housekeeping' options to reduce energy bill associated with appliances

Activity	Simple energy efficiency recommendation
Any	If at all feasible, make use of off-peak tariff(s).
Wash clothes	Use low temperature detergents. Use half load button if appropriate. Restrict cold feed pressure (if hot and cold fill).
Hob cooking	Use gas hobs or microwave ovens (where feasible). Select pan for cooking which will be at least half full. Employ well-fitting lids on pans. Turn down heat once boiling. Switch off heat before removing pan.
Oven cooking	Use as much of oven as possible per operation. Place food in oven half way through pre-heat period. If electric appliance, where feasible use toaster instead of grill and slow cooker rather than the oven.
Hot drinks	Use level gauge on kettle to boil least necessary amount. If automatic kettle, boil once and use immediately.
Illumination	Use fluorescent, not incandescent.
Hot water	Lag DHW tank. Ensure thermostat setting is compatible with providing very hot water(~ 60 °C) but not extremely hot water (~ 80 °C). Avoid making low volume (e.g. <1 litre) draw-offs from the hot-water taps. Shower, don't bathe. Energize immersion heater during 1–2 hours prior to peak use, do not operate continuously.
Television	Switch it off!

p.a. for a very well equipped home. If the household relies entirely on an immersion heater, an additional annual saving of $\sim £45$ should be achievable. It is estimated that £10–100 per annum could be saved by good housekeeping practices with existing equipment, but the inertia to modify appliance-use habits in line with energy efficiency advice should not be underestimated (e.g. refer to Table 4.7).

Most households have an array of appliances, but spend little time thinking about their energy use implications and have little idea of how much they cost to run. Yet the actions of end-users at the micro level usually influence considerably the cost of running each appliance; individual households with various stocks of appliances (Table 4.4) may find that their associated energy bills lie far outside those stated, because the analysis here assumes average usage/consumption (Tables 4.2, 4.3 and 4.4).

Clearly, those who spend least on energy for running appliances will have the least potential for reducing their financial expenditure (Table 4.4). Nevertheless, for a range of equipment levels, the predicted savings amount to 32–52% of current annual expenditures on running appliances (Table 4.6). An average non-centrally heated home employing only instantaneous electric room heaters, which wishes to raise its mean indoor air temperature to that achieved by an average centrally heated household (Table 4.2, note (c)), is predicted to require an additional 1360 kWh of heat. This will cost approximately £109 if achieved by the same form of heating – this amount may increase substantially if the dwelling is of worse than average thermal design. Although there are cheaper ways of buying heat, the option of simply increasing the use of existing room heaters may be favoured as it probably requires no immediate capital spend. In this context, the predicted savings in appliance energy expenditures (Table 4.6) are significant.

Modifications to a household's appliance stock will require capital investment in order to realize running-cost savings. Simple pay-back periods will range from a few months (e.g. to buy a timer and to shift say 33% of immersion heater use to an off-peak tariff) to several years (e.g. to replace the existing fridge with a high efficiency version). However, these periods are shortened considerably if the household already requires (or is committed to buying) a new appliance, as only the additional first cost will need to be afforded. Also, once installed, appliance energy expenditures per unit time will decrease in real terms, thereby encouraging the household to invest in buying more warmth or building improvements. Nevertheless, it will often be difficult for fuel-poor households to afford the 'replace existing equipment' or 'change to an off-peak tariff and buy timers' options of the type listed in Table 4.5. Therefore the provision of 'best practice' information and advice which is focused on how to curtail annual energy expenditures on appliances is highly desirable. If specific advice/information is to be made available to consumers, the energy utilization of domestic appliances on a micro scale requires further investigation.

4.7
Policy
recommendations

The following recommendations are put forward:

1. Establish a mandatory energy labelling scheme for all major gas and electrical appliances.
2. Enforce minimum efficiency thresholds for each category of major appliances (revised on a time base of, say, 5 years) and provide incentives for manufacturers to produce high efficiency equipment.
3. Establish one or more independent test laboratories for measuring, approving, collating and publishing the energy performances of all major appliances to be sold in the UK market-place.
4. Promulgate more energy effective appliance-use behaviour.
5. Promote independent research and development in the field of domestic energy and affordable warmth.

4.7.1 ENERGY LABELLING SCHEME

Provided that it is applied comprehensively within the retail sector, appliance energy labelling is a useful mechanism for:

- prompting individuals to consider the energy consumptions associated with their domestic activities;
- motivating manufacturers to compete in terms of appliance efficiency; and
- encouraging retail outlets to stock more efficient appliances.

The label should appear neither technical nor vague/gimmicky; it should preferably indicate the estimated average annual energy consumption/cost and provide a relative measure of where the appliance's energy performance lies within its generic group. It is essential that the information presented is derived in a scientifically rigorous manner. Weaknesses in the quality of the labels, or in the algorithms which underpin them, will adversely affect the effectiveness of such schemes.

4.7.2 EFFICIENCY THRESHOLDS

It is unrealistic to expect a comprehensive energy labelling scheme on its own to be an effective means of ensuring that individuals actually use high efficiency appliances. Many households will continue to buy cheap, low efficiency products and pay little attention to energy labels when shopping, because the achievable savings from preferring a high efficiency version amount to only a few pounds per annum (i.e. an apparently small fraction of the appliance's capital cost). Therefore a system of incentives and penalties is required to influence:

- the appliance manufacturing industry, so that energy efficiency becomes a key element of its engineering design philosophy; and
- the average purchaser, so that they view high efficiency appliances more favourably at the points of sale.

The types of incentive and the duration for which they need to be applied are strong functions of the UK market. In Germany, it appears that the average appliance buyer's definition of quality includes high efficiency (MacKenzie, 1993). However, British consumers are relatively inhibited with respect to insisting upon high quality services/goods from organizations such as appliance manufacturers. In the USA, schemes for assisting the commercial success of high efficiency appliances have been applied for several years. For example, capital cost rebates have been offered by local utility companies as a means of demand-side management; and minimum efficiency standards have been set by state or federal legislation.

Legislation is urgently required in the UK to specify minimum acceptable efficiency values for each type of major appliance. This will help:

- to ensure that manufacturers strive to improve the efficiencies of their products;
- to accelerate the withdrawal of any particularly low efficiency versions;
- and in some cases (MacKenzie, 1993), to stop the 'dumping' of low efficiency models in Britain.

Various methods may be developed to penalize manufacturers who do not meet the prescribed standards but minimum efficiency standards should be viewed positively as a means for stimulating industrial competition in an increasingly energy-aware global market-place for appliances.

It is essential that the minimum efficiency values are revised regularly in order to prevent the usefulness of energy labels diminishing a few years after their introduction because most producers are by then meeting the original definition of 'high efficiency'. Minimum efficiency legislation is likely to be introduced in Australia, partly because many products now score 5–6 points on the 6-point energy label introduced in the mid 1980s (Wilkenfield, 1993). In the USA and Canada, the combination of time-marching minimum efficiency legislation and energy labelling for appliances is viewed as an important tool for ensuring that the domestic sector takes its share of the responsibility for curtailing national primary energy consumptions, peak electricity demands and pollutant emissions (Geller, 1987; A.J. De Hart, 1994, personal communication).

4.7.3 TEST LABORATORIES

Any action with respect to recommendations (1) and (2) requires reliable energy consumption data. This information underpins the energy labels and should also be made available in a more comprehensive form (e.g. an index similiar to that which lists the fuel consumptions of cars, or simply a catalogue of the upper quartile/decile with respect to appliance efficiency within each product category). It is particularly important that the DSS, local authorities and housing associations utilize such a service in order to ensure that the households over which they have a direct influence are (where feasible) equipped with high efficiency appliances.

4.7.4 APPLIANCE USE

Actions regarding recommendations (1), (2) or (3) may well take several years to impinge upon the lives of most fuel-poor households. For example, high efficiency appliances will not be sold in significant numbers by second-hand appliance shops (which offer used appliances typically 10–40% of their brand new prices) until at least 5 years after they are first introduced. Therefore, promoting greater awareness of appliance energy utilization/thrift among low-income households is highly desirable in order to maximize achievable savings in the short term. Furthermore, with the imposition of

VAT on domestic fuel, a political opportunity now exists to influence the appliance-use habits of all types of household by defining and disseminating pertinent 'good housekeeping' or 'best practice' information/advice. The data provided for the public must be technically sound and presented in an appropriate manner.

4.7.5 RESEARCH AND DEVELOPMENT

Matters relating to energy use in dwellings have a considerable influence upon the health and well-being of individuals and upon the effectiveness of the appliance manufacturing and energy supply industries. In the context of the relatively high proportion of fuel-poor in the UK, the need to meet impending CO_2 emission targets and the fact that in many appliance categories much more engineering R & D is required to establish higher efficiency systems, the independent research base in this area is extremely under-resourced. It is time for government agencies to acknowledge the importance of this field, for example by establishing enabling mechanisms within the research councils.

References

British Gas, consumption data concerning gas-fired tumble driers, British Gas, Holborn, London, 1994.

Consumers' Association (1991) *Which?*, March and April

Eastern Electricity (1993) *Trust us for energy efficiency* (list of E-label ratings for domestic refrigeration equipment), Wherstead, Ipswich.

Evans, R.D. and Herring, H.P.J. (1990) *Energy use and energy efficiency in the UK domestic sector up to the year 2010*, Energy Efficiency Office.

Geller, H. (1987) *Energy and economic savings from national appliance-efficiency standards*, American Council for an Energy Efficient Economy, Washington DC.

GK Marketing Services (1994) *Marketing Pocketbook* 1994, Advertising Association, NTC Publications, London, p.26.

Hedges, A. (1991) *Attitudes to energy conservation in the home*, HMSO, London.

Henderson, G. and Shorrock, L.D. (1990) *Appliance energy use in UK dwellings*, Building Research Establishment, Watford.

MacKenzie, D. (1993) Britain's fridges: too hot to handle. *New Scientist*, September 4 pp. 14–15.

Newborough, M. (1994) *Electric ovens: thermal performance enhancement*. Proceedings of the International Appliance Technical Conference, Madison, Wisconsin, May 9–11.

Newborough, M., Batty, W.J. and Probert, D. (1987) Design improvements for the ubiquitous electric toaster. *Applied Energy*, **27**, 1–52.

Newborough, M. and Probert, D. (1988) Energy-conscious design improvements for electric hobs. *Foodservice Systems*, **4**, 36–67.

Ogston S.A., Florey C. du V. and Walker, C.H.M. (1985) The Tayside infant morbidity and mortality study: effect on health of using gas for cooking. *British Medical Journal*, 290, 957–959.

Shorrock L.D., Henderson, G. and Bown, J.H.F. (1992) *Domestic energy fact file*, Building Research Establishment Report, BRE, Garston, Watford.

Stephen, F.R. (1986) *Consumption profiles for electrically-heated hot-water in houses*, Report M2030, Electricity Council Research Centre, Capenhurst, Chester.

Sterlini, P.A. (1993) Efficient use of domestic appliances. *CADDET Newsletter*, No. 3, pp. 4–5, CADDET, Sittard, The Netherlands.

Stern, P. and Aronson, E. (1984) *Energy use: the human dimension*, W.H. Freeman & Co., New York.

WHO (1990) *Air quality guidelines for Europe*, World Health Organisation Regional Publications, European Series, No. 23, Copenhagen.

Wilkenfield, G. (1993) Australia's approach to improving efficiency of appliances, *CADDET Newsletter*, No. 3, pp. 6–9, CADDET, Sittard, The Netherlands.

Identifying policy objectives 5

Ian Cooper

Two major problems confronted the Group in its attempt to tackle the issues of domestic energy consumption and affordable warmth:

- choosing between policy options; and
- defining its own priorities so as to reach a consensus about what can and should be done.

Before discussing the policy options considered by the Group, it is instructive to consider, if only briefly, the framework of ideas and objectives within which energy policy has operated in the UK over the past two decades since the first 'energy crisis', especially as this relates to domestic energy consumption. Examination of the shifts that have occurred in this framework has helped the Group to identify the assumptions that continue to constrain policy objectives in this area and the current levers for change that could be used to reduce the incidence of underheating in at-risk households.

Government interest in controlling energy use is often traced back to the 'energy crises' of the 1970s, although concern about reducing energy consumption has been recurrent, in both peace and wartime, throughout this century. In the mid 1970s, **energy conservation** became synonymous with the government's 'Save it' campaign which exhorted people to use less energy even if this detracted from their lifestyles. This campaign was inspired primarily by worries about the security of energy supplies and about the impact of energy imports on the UK's balance of payments. Behind these lay a more environmentalist concern – the finite nature and the projected rate of depletion of fossil fuels. Energy conservation was generally sold then as a short-term, interim solution, a technical fix to bridge the so-called 'energy gap' until alternative, secure sources of supply became available. Whether these were seen as arising from an expansion of nuclear power or from increased exploitation of renewable energy sources largely depended, as it still does, on the ideological disposition of the commentators in question. In both the domestic and non-domestic sectors, and under both Labour and Conservative governments, energy conservation has predominantly been expected to occur through the operation of 'the market', specifically through the price mechanism as a stimulant to economic individualism. It has been assumed that, if price signals are right, then individuals (whether alone or together as households, private companies, local authorities or housing associations) will be galvanised to protect their own interests. Market forces will not only optimize how energy is produced, allocated and consumed: they will also ensure that those measures necessary to conserve it will be taken. Individuals are seen as the mainspring of action on energy conservation. Their rôle is to have sufficient surplus capital to invest in those energy saving measures which energy prices indicate, from time to time, to be cost-effective. Those without adequate disposable income to make these investments have, by definition, not found themselves included in this process. Low-income households have thus been outside the mainstream thrust of government policy on energy conservation.

Over the past 20 years, governments have (ideally) wished to see their contribution to energy conservation restricted to:

- structuring the energy 'market' so that energy prices give the right signals; and
- providing individuals with information and advice about which technical measures for saving energy are currently 'economic'.

With few exceptions, e.g. briefly under Labour (Callaghan, 1977), governments have only chosen to deviate from this preferred position to intervene in energy conservation when driven to do so by concerns that lie primarily outside 'energy policy' itself. Of course, there have been exceptions to this policy of non-intervention, for example in the form of the Home Energy Efficiency Scheme and other initiatives as described in Chapter 3. But these have, typically, been quite limited in scope and duration. They have also been piecemeal and so have not constituted a comprehensive attack on the problem of fuel poverty in the UK.

With the change of government in 1979, Britain's energy 'problem' was redefined. The late 1970s had illustrated that some of our international competitors, notably West Germany and Japan, had successfully raised their Gross National Product without increasing their national energy consumption. Britain's problem was cast as being a question of our international competitiveness, of our comparative 'inefficiency'. The Department of Energy stopped talking about a need to save energy: instead we had to use it more efficiently. Under the new banner of '**energy efficiency**', it was acknowledged that our national consumption was likely to grow. This was presented as acceptable as long as increased consumption made us more efficient, more highly productive, or delivered an increased level of service – above all, as long as it led to an improvement in our international competitiveness.

Although the phrase 'energy efficiency' was originally invented to deal with energy consumption in industry and commerce, it came to replace 'energy conservation' in the domestic sector too. Energy efficiency – i.e. the cost effectiveness of a measure as defined solely in terms of its payback period – ruled as the guiding principle by which any proposed action to reduce energy consumption should be judged.

By the end of the 1980s, because of growing concern over global warming and ozone depletion, the nature of the UK's energy problem was redefined yet again. Now the consumption of fossil fuels was identified as heavily implicated in environmental pollution, especially in the emission of carbon dioxide. There was renewed interest in reducing energy consumption in absolute, as opposed to just relative, terms. Once again energy became an environmental issue – not because of the finite nature of fossil fuels but because of what was now perceived as the rapid rate of damage being inflicted on our planet's finely balanced ecosystems by our consumption of fuels.

5.2.1 THE CURRENT POLICY POSITION IN THE UK

Discussing energy policy in the UK since the privatization of gas and electricity, Professors Fells and Lucas (1991) wrote:

> There is already a covert Energy Policy, but it is a hidden ramshackle set of checks and balances which are constantly being tinkered with ... piecemeal and even covertly, to avoid open conflict with [the Government's] political ideology that market forces can achieve all necessary results. The market is not operating freely, but even if it were it is clear that it would be in itself insufficient ... The last White Paper on Fuel Policy was published in 1967 when gas had only just been discovered in the North Sea, oil not even found, the UK had not yet joined the European Community and no-one had heard of the global environment. Much has changed, and a new policy to guide us safely into the next century must be prepared.

In the absence of this explicit policy statement, one of the particular difficulties in dealing with energy consumption in domestic premises, particularly in terms of making warmth affordable to low income households, is that no one government department possesses sole responsibility for action in this area – or, indeed, for dealing with the consequences of inaction. Even just within one department the picture is very complicated. For instance, the Department of the Environment has a wide range of responsibilities which impinge on the state of housing in England and Wales (but the Scottish Office is responsible for housing in Scotland). The 'mission statement' of the DoE's Energy Efficiency Office (DoE, 1994) is:

... to protect the environment, and save money, by encouraging better management methods and by promoting the cost-effective use of energy in all locations.

It has a UK-wide remit covering the energy performance of housing and the DoE's Global Atmosphere Division is charged with reducing the emission of CO_2 from the UK building stock. Through its Construction Sponsorship Directorate, this Department is also responsible for compiling figures on additions to the housing stock as a whole but surveys of the condition of the stock are the responsibility of the Department's Housing Division. In addition, the Department's Property and Buildings Directorate is responsible, through the Buildings Regulations, for ensuring compliance with statutory standards for the energy efficiency of any new houses built or extensions made to existing ones in England and Wales. Scotland has its own regulations which are the responsibility of the Building Directorate of the Scottish Office.

John McAllion, MP for Dundee East, tabled a private member's Bill on Building Conversion and Energy Conservation which sought to extend the energy efficiency requirements of the Building Regulations to cover all changes in use and conversions of property for residential purposes. This Bill was talked out in the House of Commons on 15 April 1994 with no date set for further debate. The DoE has provided the Association for the Conservation of Energy with a written undertaking that its provisions will be included in the 1994 revisions to Part L of the Building Regulations (Warren, 1994). Beyond this limited step, however, there appears to be no consideration at present of any further statutory requirement for the retrospective upgrading of the domestic stock as a whole, even piecemeal over time. Scotland's Building Regulations already apply to improvements to existing homes, although local authority discretionary powers and general exemptions mean that energy efficiency is not required in practice.

Other departments have different responsibilities which affect, or are affected by, domestic energy consumption and the affordability of warmth. The Department of Social Security is responsible for paying income support to those in receipt of supplementary benefits and this contains a notional, if unspecified, element to cover heating costs. The Department of Health picks up the cost of ill health caused by underheating. Less directly, the Department of Employment, the Department of Trade and Industry and the Treasury all have interests in this area because of:

- the impact that a major programme of energy upgrading of the domestic stock would have on unemployment rates, training requirements and manpower planning;
- the balance of payments (through the importation of design skills, contracting capacity and building materials if these were not supplied by UK practices, contractors or manufacturers); and
- the impact of any spending programme on the public sector borrowing requirement.

Consequently, any comprehensive attempt to reduce the energy inefficiency of the UK housing stock would have to be developed as a set of multi-departmental priorities, with co-ordinated objectives and an integrated *modus operandi* not just within but across government departments.

As a result of queries raised on this front at its one-day seminar on this topic in October 1993, the Group enquired, through the auspices of the Department of the Environment, about the current state of government policies for tackling domestic energy consumption, particularly as it relates to affordable warmth. It sought answers to ten questions concerned with both current policy commitments within departments and about integration of these across departments.

1. Does the Department of the Environment have a public statement of its policy objectives for improving the energy efficiency of the UK's existing housing stock and, if so, what rate of improvement is being sought over what period of time?
2. Apart from the Home Energy Efficiency Scheme, are the premises occupied by low-income households targeted for priority treatment?
3. Does the DoE have a public statement of its policy objectives for alleviating fuel poverty/delivering affordable warmth in low-income households and, if so, what rate of progress is being sought over what period of time?
4. Does the DoE have a public statement of its policy objectives for reducing CO_2 emissions caused by energy consumption in low-income households and, if so, what rate of reduction is being sought over what period of time?
5. What mechanisms does the DoE intend to employ to achieve these reductions?
6. Given the DoE's 1992 discussion paper on *Climate Change*, what proportion of the CO_2 reduction that the government is seeking from the domestic sector is being sought by:
   • reducing energy consumption
   • switching fuels
   • increasing energy efficiency
   specifically in premises occupied by low income households in the UK?
7. Does the Department of Health have a public statement of its policy objectives for reducing expenditure caused by ill health arising from inadequate heating in low income households and, if so, at what rate is this ill health to be reduced over what period of time?
8. When did the Department of Health last discuss its policy objectives on this front with the DoE and were any joint objectives agreed at that meeting?
9. When did the Department of Social Security last clarify what proportion of income support to low-income households is intended to alleviate the hardship caused by underheating?

10. When did the Department of Social Security last discuss its policy objectives on this front with the DoE and were any joint objectives agreed at that meeting?

The Group has received no formal replies to these questions. Given this lack of response, any recommendations made by the Watt Committee arising from the Group's report are in danger of falling into what may be described as a policy vacuum on this front in the UK.

The Select Committee on the Environment, in its Fourth Report on *Energy Efficiency in Buildings* (1993) recommended that:

> ... the Government seek to co-ordinate across its various Departments those programmes and campaigns that relate to energy efficiency. An intra- and inter-Departmental committee of key officials may be one method for achieving this.

In its response (DoE, 1994) to this recommendation, the Government made reference to meetings of its 'green' Ministers on energy efficiency and explicitly to a need for collaboration between the Department of the Environment and the Department of Trade and Industry. No reference was made to any requirement for co-ordination between these and the responsibilities and programmes of the Departments of Health, Social Security and Employment.

Given the government's inaction on affordable warmth in the UK, it is possible that the next move may come from outside, from the European Union where DG 17 has recently expressed interest in 'fuel poverty' in the UK.

5.2.2 CURRENT LEVERS FOR CHANGE

The first potential lever for change within the UK is the government's own response to climate change. It has given a conditional commitment to return the UK's emissions of carbon dioxide to the 1990 level by the year 2000. Three policy initiatives have been introduced to achieve this:

- a UK carbon strategy (currently shelved);
- a formation of an Energy Saving Trust; and
- the introduction of incentives for the gas and electricity industries to encourage energy saving.

These have been described (Warren, 1993) as measures created to 'force' the market-place to deliver the reduction of 10 million tonnes of carbon – from 170 to 160 MtC – that is required to meet the government's stated policy objective. Perhaps just as significantly, the adoption of a policy on global warming has two other consequences:

- Apart from fuel substitution (and hence the 'dash for gas' as a more environmentally friendly fuel), it commits the government to securing real (i.e. absolute) reductions in energy consumption rather than, as previously, simply relative increases in the efficiency with which an ever-increasing amount of energy is consumed.

- It provides an explicit benchmark against which to measure the success of the government's implementation of its policy objective over time, and of the practices it has employed to achieve this – an external reference point which was lacking before.

The Global Atmosphere Division of the Department of the Environment is now charged with delivering an absolute reduction in energy consumption in the UK because of its impact on CO_2 emissions. The DoE is seeking to reduce the 26% of carbon dioxide (41 MtC) currently emitted by UK households, as it made clear in its discussion document, *Our National Programme for CO$_2$ Emissions*. But it is impossible to gauge from this document how much of the reduction in CO_2 emissions that the government is seeking from the domestic sector is expected to come from reducing energy consumption in low-income households. Given the emphasis in the document on *Voluntary Action and the Individual Citizen*, probably little is anticipated on this front. There is a real anomaly here, given the DoE's launching of its *Helping the Earth Begins at Home* campaign. Presumably the only households being exhorted to act here, at least in terms of making their homes more energy efficient, are those with surplus income available for investment. Once again, low-income households seem to fall outside the thrust of government policy.

The Energy Saving Trust may also prove to be influential. In November 1992, the government launched the EST with a remit to develop and manage a new programme to promote the efficient use of energy in the domestic sector. The Trust is a joint initiative between the government and the fuel utilities, British Gas and the regional electricity companies in England, Scotland and Wales. By 1999, it was proposed that the EST should be spending £350–400 million a year on energy conservation projects.

In the spring of 1993, the Chancellor of the Exchequer announced the imposition of VAT on domestic fuel. This was subsequently signalled as a measure in line with the government's commitment to reducing CO_2 emissions. There was a public and parliamentary outcry at the placing of this additional financial burden on low-income households – it is to be applied not only to fuel consumed but standing charges too. The Chancellor responded in November 1993 with two compensation packages, running for a three-year period and rising to £400 million. One is aimed at all pensioners, the other at low-income households. The former are to receive 75% and the latter 50% compensation for their additonal expenditure on fuel bills. In addition, the government has allocated a one-off addition of £35 million to HEES to cover all pensioners. In the longer run, it may find itself required to accept that no further economic measures can be introduced in Britain to restrain energy consumption, such as the proposed carbon/energy tax, until the issue of fuel poverty has been tackled. Eventually, it may have to concede the re-introduction of explicit DSS income support for heating benefit. However, unless this is followed by a capital investment programme to reduce energy consumption in low-income households, the government will simply be committing itself to additional heating payments *ad infinitum*, to an unending subsidy of energy inefficiency.

It is here that the EST has looked, until recently, the most immediately hopeful lever for change. Currently its remit is specifically focused on energy use in the domestic sector and one of its programmes does cover increasing energy efficiency in social housing. More importantly, its source of funding, directly from the utility companies, is highly advantageous given the government's desire to reduce public spending, for the Treasury is unlikely to accept any new initiative to reduce the incidence of underheating whose funding adds explicitly to the public sector borrowing requirement. However, the role of the EST as a possible lever for change here has been thrown into doubt by the evidence presented to the Environment Select Committee by the new Director General of Ofgas, Clare Spottiswoode, who was appointed in November 1993. After taking legal advice, she informed the Environment Select Committee in March 1994 that the current arrangements for funding the EST used by British Gas are 'a tax raising authority' that go 'beyond the powers I have been given under the Gas Act'. This has been described by the Chairman of the Select Committee, Robert Jones, as 'driving a coach and horses through the entire energy saving strategy of the UK'.

More recently, the Energy Conservation Bill (1993), a private member's bill adopted by Alan Beith MP, the Deputy Leader of the Liberal Democrats, which had been accepted by the House of Commons without a vote, passed on to its report stage (April 22 1994) where it was killed by a filibuster from the government's Energy Minister, Tony Baldry. If this Bill had been enacted, it would have required local authorities to conduct an energy audit of the housing in their area, both public and private sector, and, on the basis of the information collected, to draw up a local energy conservation plan. These plans, which were to have been drawn up in consultation with all relevant local interests – Chambers of Commerce and Trade, consumer and environmental organizations, parish councils and other community groups – would have required local authorities to estimate the cost of achieving energy savings of 10, 20 and 30% (ACE, 1994). The Bill had been endorsed (ACE, 1993) by Sir William Doughty, the President of the Association for the Conservation of Energy, precisely because:

> ... by tackling the energy efficiency of individual homes, [it] addresses the cause of cold, damp homes and not just the symptoms.

On defeat of the Bill, for which there had been all-party support within the House of Commons, Alan Beith commented (Mintour, 1994):

> This government now has no energy conservation policy – and no commitment to helping the poor, disabled and elderly people to make their homes energy efficient.

A major task facing the Group was to fix the scope of what it was trying to achieve. Should it pursue the adoption a single, fixed objective, e.g. the alleviation of all underheating, regardless of its impact on other existing areas of government policy? Or should it adopt a broader set of objectives based on compromise with existing policy objectives, e.g. assisting at-risk households via the notion of 'best achievable fit' between:

- alleviating underheating;
- improving energy efficiency; and
- reducing CO_2 emissions for domestic premises?

**5.3
Policy options
considered by
the Group**

The Group identified at least four distinct policy options which could be adopted.

1. A programme intended to produce (a phased) alleviation of underheating for all households 'at risk'.
2. A programme intended to alleviate underheating for households 'at risk' only where this is compatible with the application of traditional 'cost-effective' criteria for energy efficiency measures.
3. A programme intended to alleviate underheating for households 'at risk' only where this is necessary to achieve the target specified for the government's conditional commitment to reducing CO_2 emissions.
4. A programme of work intended to benefit 'at risk' households based on a 'best fit' compromise between three objectives – alleviating underheating for 'at risk' households, increasing the energy efficiency of their accommodation, and reducing its emissions of CO_2.

**5.4
Policy option
selected by
the Group**

After protracted discussion, the Group has agreed that only the first of these policy options would, of necessity, alleviate underheating for all households 'at risk'. Such a programme of work would, realistically, have to be phased since it could only be funded over a period of time. However, if it were to be phased, criteria would need to be developed for deciding the order in which cases of underheating would be treated.

Whilst one of the by-products of such a programme of work would be to improve the overall energy efficiency of the UK housing stock, it should not be assumed that, in the first instance at least, this would necessarily lead to a decrease in domestic energy consumption. Some households might quite properly, given the increased energy efficiency of their circumstances and the very cold houses they live in, choose to take this in increased warmth and comfort rather than reduced energy consumption. For that reason, there might, initially, be little or no reduction in CO_2 emissions from domestic premises occupied by low-income households as a result of the implementation of such a programme of work. Furthermore, in order to deliver affordable warmth to such households, it might be necessary, on occasion, to take energy efficiency measures which could not be justified in terms of simple cost-effectiveness criteria (which do not take into account, for instance, health improvements and consequent savings on health costs).

As a consequence of this sequence in its decision making, the Group's chosen policy option is aimed squarely at alleviating underheating. It is only secondarily focused on what is traditionally meant by cost-effective energy efficiency and only indirectly orientated towards reducing CO_2 emissions.

References   Association for the Conservation of Energy (1993) *The Energy Conservation Bill 1993: How it could work*, ACE, London.

Association for the Conservation of Energy (1994) *The Fifth Fuel: the newsletter of the ACE*, No. 28, Spring 1994, ACE, London.

Department of the Environment (1994) *Climate Change*, DoE, London.

Environment Select Committee (1994) *Energy Efficiency: the role of OFGAS*, Minutes of Evidence, HC 328-i, HMSO.

Department of the Environment (February 1994) The government's response to the Fourth Report from the House of Commons Select Committee on the Environment: Energy Efficiency in Buildings, CM 2453, HMSO.

Fells, I. and Lucas, N. (1991) *UK Energy Policy post-privatisation*, Scott Stern Associates, Glasgow.

Mintour, P. (1994) 'Minister talks out conservation bill', *The Guardian*, 23/4/94, p.4.

Warren, A. (1993) Energy's year of optimism, *Energy in Buildings & Industry*, Energy 93 Show Guide Catalogue, p.4, Westerman, Kent.

Warren, A. (1994) private communication, Association for the Conservation of Energy, 26 April 1994.

Strategies for action 6

Brenda Boardman

The problem of fuel poverty is getting worse: more people are in poverty and the quality of their homes is hardly improving – certainly not relative to higher income groups. The imposition of VAT on domestic fuels will deepen and extend fuel poverty, firstly because most existing sufferers cannot be compensated sufficiently through the package of measures announced in November 1993. Secondly, because the imposition of VAT has made fuel more expensive and extended the boundaries of fuel poverty, more people will have inadequate levels of warmth. There is now widespread concern about the problems of the fuel poor and a growing recognition that their homes have to be made energy efficient before further price or tax increases can be imposed. This is particularly true of the proposed European-wide carbon and energy tax.

Therefore, environmental and energy efficiency policies in the UK hinge on first reducing the problem of fuel poverty. This is the new situation that gives this Report so much added impetus.

Any programme of action has to be assembled from the answers to key decisions. Chapter 5 rehearses the Group's debate about one of these and confirms that the focus is on reducing fuel poverty; cost effectiveness and carbon dioxide reductions are considered as secondary objectives. The design of an appropriate programme for affordable warmth involves several other dimensions and the Group's thoughts on these are presented below. Chapter 7 demonstrates how these might be enacted in practice in one proposed scheme.

It is important to recognize that different responses to the following questions would result in changes of strategy. The Group's responses are not the only, or necessarily the best, solutions. They are, however, the collective view of a concerned group of energy and fuel poverty specialists who have met on 12 occasions over 2 years. The Group hopes, therefore, that they will be seriously considered as the basis for policy action.

6.1 Providing affordable warmth

The average low-income household in receipt of a means-tested benefit will only have affordable warmth if its dwelling has a level of energy efficiency equivalent to a National Home Energy Rating (NHER) of 8 or higher. This standard will enable the levels of heating identified in Chapter 2, together with equivalent other energy services, to be obtained for a maximum weekly expenditure of 10% of household income for all fuel, including standing charges. It is acknowledged that this is a high standard – above that achieved by a new dwelling complying with the Building Regulations – but it is dictated by the level of income received by these claimants.

6.2
Time-scale

There are now 8 million households in receipt of a means-tested benefit or disability allowance and, with a few exceptions, suffering from some degree of fuel poverty. The lifetime of many insulation materials and heating systems means that each house has to be revisited and retreated on a 15–16-year cycle. After this time, the performance of the materials or heating equipment can no longer be guaranteed. (The main exception is draughtproofing materials which are durable for only 5–10 years, if good quality, and need replacing on a shorter cycle.) Therefore, all the fuel poor have to be provided with affordable warmth within 15–16 years.

6.2.1 STRATEGY CHOICE

(a) Upgrade each house once every 15–16 years and make sure that it has achieved a standard of affordable warmth at the first and only visit, at a rate of just over 500 000 p.a. (It is assumed that half a million a year is about the maximum that could be undertaken, because of the administrative implications and limits of available skills and funding.) As poverty is increasing, this means that a programme for affordable warmth would now take 16 years rather than the 15 years anticipated in earlier publications.
(b) OR: Visit each house several times and gradually improve the quality of all the low-income housing stock, perhaps dealing with 3–4 million properties each year and visiting each house up to six times over the whole 16-year period. (The maximum of six visits is based on research by Newark and Sherwood District Council amongst their tenants about the level of disruption that is acceptable.)

6.2.2 RECOMMENDED APPROACH

Some combination of both, as the local agency, the occupant and the home owner choose. The main examples of a major or total upgrade can occur when:

- a new property is built, for a low-income family;
- an unfit property is made fit for habitation, if discretionary grants are included;
- a property becomes empty and is given a major retrofit by the landlord.

In all other cases, the work will probably be undertaken on a piecemeal basis, mainly to avoid disruption to the occupant. A range of options already exists (Chapter 3), which could be increased in scale. For instance:

- HEES could include new measures, such as cavity wall insulation, fitting heating and water controls and repairs to appliances.
- HEES should be enlarged if 8 million means-tested claimants are to have their homes draughtproofed every 5 years. This would mean a six-fold increase in the present level of work on low-income properties, from 240 000 p.a. to 1.6 million p.a.
- Local authorities could be required to make the House Renovation Grants for thermal insulation available on a specified basis, rather than discretionary.

The scale of work and the budget should be sufficient to recognize that the state of fuel poverty is very demoralizing and hardship has to be relieved, even if only incrementally in each household, as soon as possible. Thus there must be an emphasis on the second approach, for equity reasons. This will mean that heating standards outlined in Chapter 2 will be obtained in stages for the majority of households. On the other hand, the temperature in the 10% of coldest homes will be getting gradually warmer each year and at a faster rate than historically, if our proposals to target the fuel poor are accepted.

The assumption is that the local authority is the only institution capable of delivering a comprehensive policy approach (Chapter 7) and that an overall target is needed to encourage and ensure action. The definition of these targets is difficult with the paucity of information available to the Group, but suggestions are:

- Raise the quality of all housing within the local authority's boundaries by an average of 1.5 points of NHER in 7 years. The average in the UK is just below 4.5 now, so by the end of 14 years it should be up to 7.5, close to the standard that provides affordable warmth for low-income households.
- To protect the poor, a parallel requirement is that the number of homes below NHER of 2 has decreased by 50% in the first 7 years and that there are none by the end of 14 years.

The accurate and careful setting of these targets will provide the best assurance of delivering affordable warmth. However, the constituent schemes that deliver this level of activity will need to be supported by a substantial increase in existing budgets. In addition, the progress in each local authority will need to be carefully monitored and penalties imposed for a lack of action. The allocation of the Housing Investment Programme is already linked to audits of the stock and declared action plans. This sets a precedent for clear energy efficiency investment plans to be rewarded with additional funds. The penalty for inaction, at the moment, is a decline in that authority's budget, to the further detriment of the poor. This latter situation will need to be addressed. There is a real risk of the geographical concentration of fuel poverty in those local authorities that have large numbers of poor quality housing and/or small budgets. In addition, not all local authorities are equally innovative and, whilst it is appropriate to encourage best practice, policy has to address the problem of the average and the laggard as well.

6.3 Complete coverage

How can all houses that need to be treated be identified? And how can the level of treatment be monitored? Many of the present schemes, such as HEES, depend upon self-identification: the individual householder has to take action and approach the insulation installer. Only rarely will one of the caring professions liaise between the needy and the energy efficiency provider. There is no database of who has been helped and certainly no way of identifying the worst properties and tracking whether they have been made easier to heat.

6.3.1 STRATEGY CHOICE

(a) Should the focus be solely on benefit claimants? There is a risk of rejection because having the work done identifies them as 'poor' and they do not want this stigma, and there are a lot of eligible non-claimants and the nearly-poor, who should not be excluded.

(b) OR: Tackle a wide definition of poor households and poor housing. How would the Group define the limits and ensure that all appropriate households (people and buildings) had been included? Would this also need a new means-test? The Group certainly recognizes that it is dealing with both poverty and housing parameters and there is an uncertain overlap.

(c) OR: Should the programme identify only the energy inefficient properties through an audit of all houses, regardless of the income group of the occupant?

6.3.2 RECOMMENDED APPROACH

The Group's primary focus is the fuel poor – the poorest households in the worst dwellings. It accepts that the problem of inadequate warmth extends substantially beyond this group and wants policies to include those that are 'nearly poor'. The boundaries of fuel poverty are changing, for instance through the impact of VAT and, probably, as a result of real price increases as the domestic gas market becomes competitive. Also, the population is mobile and people move to different addresses. In addition, claimants are dispersed in the housing stock, perhaps occupying 50% of a particular block or estate. For technical and economic reasons, it is often appropriate to improve the whole area at the same time. Therefore an extensive identification process is needed that will cover more homes than just those presently occupied by claimants.

The first stage of any programme for affordable warmth will be to establish a strategy to identify the worst properties and to allow subsequent monitoring of progress. This can be separated from the responsibility to undertake the work. Whatever data are used, the process of identification must be mandatory, to ensure that every property is included. One possible approach is for Council Tax bands to recognize the level of energy efficiency, though the Group prefers separate energy audits (discussed below). A financial link is appropriate so that householders and landlords obtain a monetary return for investing in energy efficiency. This would come from either reduced Council Tax or increased rents. Chapter 7 examines one methodology in detail.

**6.4
Energy
audits**

Energy audits are discussed in several places in this Report. The Group stresses the importance of an audited approach because this enables the physical parameters (the house, heating system, appliances and floor area) to be combined with economic factors (standing charges and unit fuel costs), as well as the social ones (numbers and ages of people in the household). The Group believes that the geographical variations in climate are also relevant. The National Home Energy Rating includes all these variables; the Government's Standard Assessment Procedure (SAP) and the Starpoint scheme are more restricted. Perhaps the most crucial inclusion is the cost of the fuel. The energy efficiency of the dwelling can be measured solely in physical units, but this is of limited usefulness when discussing pay-back periods and policies for different income groups. Audits can be used as the basis for:

- setting a desired standard – either as a minimum (e.g. unfit for human habitation because it is too expensive to heat) or to be achieved (as in the proposed Building Regulations);
- identifying those existing properties that fail to meet a specified standard;
- determining the relative benefits of measures within a particular dwelling;
- paying additional benefit to claimants to compensate households for VAT imposition and for variations in energy efficiency between different claimants on similar incomes;
- setting rent and rate levels.

None of the audits that presently exist provide outputs to satisfy all the above functions. The last two require actual expenditure, rather than an index. At the moment, the audits add up the costs of providing a specified level of service and then divide by the floor area, to give, effectively, a rate of energy consumption per square metre. This is excellent for comparing houses, but to ensure benefit adequacy or set rents it is also necessary to know the total expenditure needed.

6.4.1 STRATEGY CHOICE

(a) What function for audits?

(b) AND: What level of accuracy? There are varying levels of sophistication with audits, from walk-throughs by trained personnel, to self-administered paper questionnaires or computerised audits, such as NHER.

Anything detailed would cost a lot of money. Even if the cost per house could be brought down to £20, the total bill for the 8 million properties occupied by means-tested benefit claimants becomes £160 million. The number to be coded is likely to be wider than just the 8 million to include the 'nearly poor', however defined.

(c) AND: What time-scale for implementation? The more sophisticated it is, the longer and more expensive it is to introduce.

6.4.2 RECOMMENDED APPROACH

The Group strongly believes in the benefits that come from an approach that utilizes audits, as this provides a fair and equitable method of assessing properties, work and assistance. Public opinion appears to support policies that take action on the 'worst' housing as defined by audits. However, the accuracy of the audit (and therefore its cost) can vary. For instance, initially, to identify the worst housing, a relatively simple approach can be used, as typified by Level 0 (or 0+ as defined in Chapter 7). This is based largely on existing data held by the local authority and looks at blocks, estates or areas. It is not address-specific. A level 0 approach would be sufficient to draw up a stock profile (of all housing, in all tenures) to determine the numbers of properties that are unfit, below an NHER of 2, or similar standards necessary for monitoring progress and grants.

Once the areas of the worst properties have been identified, a more detailed survey can be undertaken to fulfil the other objectives listed above. Again the audit should provide only the level of detail actually needed.

This combined approach would have the benefit of both limiting expenditure on audits and of speeding up the process of auditing, whilst still providing the information necessary for planning.

The Group is particularly disappointed that the Energy Conservation Bill, proposed by Alan Beith, was talked out in Parliament. The Bill would have provided the right legislative basis for the proposals in this Report and its failure demonstrates how far the UK is from developing a strategy to assist the fuel poor.

**6.5
Additional
Social
Security
benefits**

Any capital investment programme for affordable warmth will take 16 years to implement. Millions of people are suffering from fuel poverty meanwhile. This hardship will be increased by VAT and by the possible introduction of a carbon and energy tax.

6.5.1 STRATEGY CHOICE

(a) What should be the balance between providing short-term amelioration through additional income and the longer-term solution of capital investment? Does the Group suggest that additional benefits are necessary, even though this increases energy use, pollution and government expenditure in the short term?

(b) OR: Accept that the administrative complexity of targeting properly and substantial cost to government make additional income an unlikely policy and push for all expenditure to be on actual improvements?

(c) OR: Propose that both expenditures are necessary initially and that the additional income can only be reduced as the capital investment is effective?

6.5.2 RECOMMENDED APPROACH

The Group's primary objective is the relief of fuel poverty (Chapter 5). Therefore, a supplementary income payment to enable the 8 million claimants to have affordable warmth in their present homes is needed now. This is recommended even though it has negative impacts on other policies in the short term, by increasing government expenditure and creating additional pollution. The DSS has agreed (in private discussions) that it would add an energy efficiency code into the benefit system provided that it was not involved in the expense and administration of identifying the code. Thus, procedurally, the payment is relatively simple once the audit code has been obtained. An easy first step to coding would be to build on the regional approach suggested by the Campaign for Cold Weather Credits. At the moment the benefit system does not reflect any regional variation, for instance because of differences in climate.

With recipients of housing benefit, the additional income should be provided as quickly as possible. The impact of the second phase of VAT in April 1995 will result in an even greater discrepancy between the increased costs and the compensation provided, and even the latter ceases after two years. The need is growing for a proper assessment of heating and energy costs, per claimant, and for the provision of income sufficient for adequate warmth to be a choice in these families.

With recipients of housing benefit, the effect of the code would be to alter the scale of the needs allowance: claimants in the inefficient homes are understood to need more money to live on, so that they get greater assistance with their housing costs. As Chapter 2 explains, this new heating benefit could cost £4 billion per annum and result in the emission of an extra 13 MtC.

The capital investment programme, as explained above, has to reach all 8 million homes in a maximum of 16 years. Any extension of the initial programme for affordable warmth will mean that it overlaps with the second programme of visits to some homes, where the standard of equipment has to be reassessed. The result would be an increasing, rather than decreasing, workload.

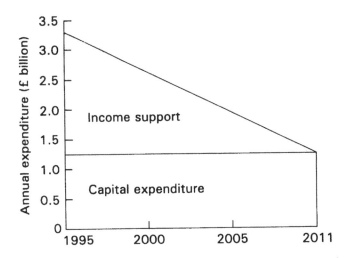

Figure 6.1 Government expenditure profile to provide affordable warmth in the UK.

It is not possible, therefore, to have a fixed level of expenditure (like a repayment mortgage) that is divided between income support and capital investment. The size of the problem and the hardship created dictate that both income support and capital investment need to start immediately. The total cost will be highest at the beginning, but will taper off as the capital expenditure reduces the need for the heating benefit (Figure 6.1).

6.6 Regulations

Regulation costs nothing and forces non-governmental agencies to make the expenditure, though this may feed back into government expenditure. Whilst UK policy is opposed to regulation, as it interferes with the market and creates 'red tape', some regulations with an energy efficiency component exist or are being discussed:

- New Building Regulations, based on energy audits. Considerable discussion has been taking place as to whether the Building Regulations should be concerned only with the building fabric (which lasts the longest) or with everything used in the dwelling, e.g. lighting, heating and hot water, in a combined energy audit.
- The extent to which Building Regulations can be applied to conversions of existing properties.
- Linking the fitness standard to an energy audit based on providing affordable warmth. This would have implications for local authorities as the work is covered by a mandatory grant.

- Minimum standards of energy efficiency for electrical appliances. These are gradually being introduced through European Union directives. For instance, the cold appliances (fridges and freezers) will have minimum standards set in 1994, introduced by 1997, with further improvements by 2000. The wet appliances (washing machines, driers and dishwashers) will have a similar, but slightly lagged, timescale.
- A minimum standard of energy efficiency for all rented property. This has no official status yet. The fear is that this will require the introduction of a registration scheme and, perhaps, a return to 'fair rents' and will cause a decline in privately rented accommodation. A counter-argument is that market forces will deliver more energy efficient housing, as it is easier to let. The Group largely rejects this latter perspective, as there is too much poverty and need for cheap housing.

6.6.1 STRATEGY CHOICE

(a) Does the Group propose the use of supportive regulation, despite official reluctance, because it helps to set both minimum and desired standards at minimal cost to government?

(b) OR: Should reliance on regulations be limited, because of the implied need for additional regulatory bureaucracy and because it is too slow to influence the present situation, especially for the fuel poor?

6.6.2 RECOMMENDED APPROACH

The regulatory approach is used wherever possible, in the interests of energy efficiency and energy conservation in the whole domestic sector. This will raise standards and expectations, but only slowly, so the 'trickle-down' effect for the fuel poor will be minimal. It will be important, however, to ensure that any new regulatory standards that primarily benefit the better off are counter balanced by measures specifically to aid low-income households. For instance, improved Building Regulations standards for new homes mainly benefit higher-income households, so a parallel set of improvements is needed to upgrade the energy aspects of the fitness standard if the present spread of thermal efficiency is not to be made even wider. This dual approach of improving the best and the worst together has not occurred previously. Similarly, the poor must be provided with access to more efficient appliances, whether this is through the Energy Saving Trust or the Social Fund.

The worst scenario would be for improved standards, backed by regulation, to benefit solely the better off and thus contribute to the further relative disadvantage of the poor.

6.7 Energy efficiency bias

Is there a danger of an inappropriate bias in emphasizing energy issues? Financial resources are scarce and a lot of housing is unfit or lacking in amenities. Should society be concerned about energy efficiency, if people have no hot water source or have a toilet at the end of the garden, or even no home at all and are living in bed-and-breakfast accommodation or on the streets? Many housing specialists and civil servants believe that energy efficiency is a second order priority and that the primary need is a decent home for all.

6.7.1 STRATEGY CHOICE

(a) The problems of homelessness and unfit properties are so great that the Group cannot justify asking for resources for energy efficiency.
(b) OR: Housing strategies should include the cost of bringing the housing stock up to a standard that recognizes adequate warmth is an integral part of proper shelter.

6.7.2 RECOMMENDED APPROACH

All the other housing difficulties are acknowledged and the Group is supportive of the need to deal with them urgently. Many housing problems exacerbate a form of fuel poverty: it is cold and unhealthy to be sleeping rough, and living in an unfit house can be nearly as bad. Bed-and-breakfast accommodation often provides minimal standards for a limited number of hours a day beyond which the claimants are forced out of the building.

The deteriorating condition of the housing stock has to be recognized, as this is inexorably creating another and much larger problem. Most houses are built to last 60 years, whereas the average age of a home in the UK is now 50 years. In addition, as Chapter 3 demonstrates, the fabric of the existing stock is not being repaired to prevent deterioration and insufficient expenditure is going into energy efficiency improvements. The housing stock is not being replaced and treated at the expected rate and the sooner this situation is reversed, the better.

There is real poverty and also a legacy of poor quality housing, so the costs of achieving affordable warmth are going to be extremely high. This has to be stressed and acknowledged. The Group is now highlighting the need for an extra layer of investment that has not previously been recognized. This omission is resulting in an increasing number of homes that provide poor quality shelter, because the occupants cannot afford adequate warmth. There is limited money, but resource constraints now will result in the need for greater expenditure in later years. It is no longer possible to separate out energy efficiency from fitness. Therefore the Group proposes:

- New money to deal with energy efficiency, with an emphasis on both the integrity of the fabric and the energy efficiency improvements (because, for instance, loft insulation deteriorates under a leaking roof).
- The money going on energy-related expenditure should be used more

effectively (Chapter 3). The latter included about £350 million in capital programmes on English local authority housing only in 1991/2. (Boardman 1993, pp. 317–8). The Green House Programme demonstrated how careful assessment of schemes can result in effective expenditure.

6.8 Cost effectiveness, cost benefit or affordable warmth?

The Group has distinguished between these three situations:

- **Affordable warmth** is a standard (as described in Chapter 2) to be achieved, no matter what the cost per individual property.
- **Cost effectiveness** indicates the allocation of money to properties and groups of people and on measures in relation to the financial benefits that accrue within a household. The boundaries of the financial analysis are the individual property and its occupants: the expenditure is justified in terms of the benefits of additional warmth and saved expenditure that is experienced by the household. The definition of cost effectiveness will require careful discussion, but there has to be a pay-back in 16 years, before the second visit. The Group has also assumed that the pay-back is based on both the total energy saved and the value of the additional warmth, not just the former.
- **Cost benefit analysis** takes a wider, societal perspective. The value of the benefits to society is added to those that accrue solely within the house. For instance, the savings to the National Health Service, the value of better educated, healthier children, the advantages to landlords and private companies of reductions in rent arrears and fuel debts are all included in the value to society of an improved, energy efficient housing stock.

6.8.1 STRATEGY CHOICE:

(a) Is the objective a standard (affordable warmth)?
(b) OR: Should some form of cost effectiveness limitation be accepted that some people in some homes cannot be given affordable warmth?
(c) OR: Should a societal, cost benefit approach be accepted that the wider benefits, such as good health, uninterrupted schooling, improved housing stock, should be recognized when determining what investment should occur?

6.8.2 RECOMMENDED APPROACH

The Group's primary objective is to provide affordable warmth for all: however, it is accepted that in some cases this could only be achieved at inordinate cost. Therefore, the provision of affordable warmth for some claimants in the present housing stock is not likely to be cost effective, however that is defined. Examples of likely problems include:

- People (notably the single, under-25s) who are on particularly low incomes, whose cost of heating needs to be reduced to about £3 a week. This is not impossible, just expensive. One implication is that there should be more new housing specifically for the young and mobile (e.g. housing co-operatives) in inner city areas to replace poor quality multiple-occupied premises.
- Those elderly still occupying large family homes, where the amount of space makes it uneconomic to improve for just one person's income. Another policy implication is that there should be more housing, whether sheltered or not, specifically for this group, to intermediate, rather than minimal, space standards.
- The fabric of the property can be expensive to upgrade, for instance for that third of the housing stock with solid walls. These properties pose real dilemmas for an energy efficient housing policy: the choice is between improvements that are not traditionally cost effective (including external insulation, building a second wall to form a cavity), continuing income assistance to the occupants, or planned demolition.

The Group aims at affordable warmth; it believes that the cost-benefit approach would result in the wider recognition of the full societal costs of fuel poverty and thus release greater sums for investment. But it accepts that, in the short term, the predominant policy choice will be a cost-effective one. The problems inherent in a societal cost-benefit analysis are discussed in Chapter 5 as these seem to be largely driven by inter-departmental problems in government. This is a real disadvantage to a comprehensive approach, which would increase employment and stimulate demand in a flagging energy efficiency and construction industry.

**6.9
Administrative
costs and
economies of
scale**

Clarification will be needed with regard to administrative costs, how they are included in the analysis and what might be meant by 'economies of scale'.

The simplest schemes, such as the present HEES, undertake a single measure, or limited package of measures, every time; a further example would be a programme to install cavity wall insulation. Under these schemes, it is assumed that the measure is cost effective in each house treated. The cost effectiveness assumption may not be correct – there will be some homes where the cost of the work is only repaid by the energy savings over a long time-scale. But the scheme is administratively simple.

The administrative costs of going to each house may be isolated from the cost of the work. This is the procedure for HEES; the Energy Action Grant Agency's overheads are in addition to maximum grant levels and paid directly by Government. Another example is the Energy Saving Trust, whose overheads, at least in this initial period, are separate from the investment programme and paid by the utilities and Government.

The administrative costs (and disturbance) will be less per job if a lot of work is undertaken in a single house at one time, for instance, to bring a dwelling up to a level of affordable warmth. The inclusion or exclusion of administrative costs may be sufficient to make the difference between cost-effectiveness or not.

The other aspect is that there are significant economies of scale from doing the same measure in a group of houses (e.g. cavity wall insulation) and from the administrative saving of doing everthing to one house at the same time. The relative value of these economies will vary. Cavity wall insulation is by far the clearest example of the first, with costs per square metre coming down by at least a factor of five for bulk activity. With the second category, the size and cost of a heating installation can be reduced if it is combined with a good level of insulation.

6.10 Fuel poverty vs the environment (or warmth vs energy saving)

The Group has decided that the problems of the fuel poor must come first and that energy and carbon dioxide reductions are secondary issues. It accepts that some policies to reduce fuel poverty could further harm the environment:

- Additional income support in the short term will not be offset by energy efficiency improvements from capital investment.
- Switching a household from reliance on an expensive fuel to a cheaper one to reduce fuel poverty will increase energy consumption: the household can spend less, but obtain more fuel. In some cases, for example a change from on-peak electricity to off-peak heating, the increased consumption will also increase pollution.

In both cases, the additional pollution results from the fact that people at present are cold: they need more warmth and will purchase more when they can. When energy efficiency improvements are applied to warm homes, a greater proportion of the benefit is delivered as real energy savings and thus lower emissions. At the moment, it is accepted that across all income groups in the UK there will be an average of 70% of the benefit of an energy efficiency improvement taken as reduced energy consumption and that the remaining 30% will be taken as additional warmth (Energy Saving Trust, 1994). This average includes better-off households; for low-income households the average ratio is likely to be reversed, to about 30:70.

6.10.1 STRATEGY CHOICE

What is the time-scale for the conflict between fuel poverty reduction and environmental gain and can this be ameliorated?

6.10.2 RECOMMENDED APPROACH

The best way to reduce the impact on the environment is to undertake a substantial programme for affordable warmth as quickly as possible: that is, to get homes warm in the short term, through additional income, and then make sure the capital investment follows quickly. In that way, the extra environmental pollution will be short-term and the subsequent energy efficiency improvements will have even greater savings.

6.11
Occupant
vs owner

New energy efficiency measures are increasingly going to 'interfere' with the building fabric and need the permission of the building's owner before they can be implemented. Present policies and much of the Group's concern focus on the building's occupant.

6.11.1 STRATEGY CHOICE

(a) Should the focus be on the building's owner (regardless of income) and provide subsidized energy efficiency improvements, just because the occupant has a low income?

(b) OR: Is the 'rich' landlord (including local authorities) required to pay for the work, in which case how can they be forced to have the work done – through minimum regulatory requirements? How can an increase in rent be prevented, or should one be allowed within limits and where the fuel savings will offset the additional rent?

6.11.2 RECOMMENDED APPROACH

Even with basic measures, such as those available through HEES, the landlord's permission is needed. With local authorities, the permission is usually readily given, as it is for large blocks of property owned by private landlords. The need to convince small private landlords of the calibre of the installer and to obtain permission to do the work, sometimes on a single property, is one of the reasons for the restricted level of HEES activity in the privately rented sector. The Group accepts the need for landlords to be involved and recognizes the problems that this causes in the privately rented sector. The administrative costs would be reduced if landlords had an obligation to provide housing with a minimal level of efficiency.

It is likely that a greater financial contribution can be expected from public rather than private landlords if the work is to be undertaken. Both sets of landlords may be encouraged to invest if the improvement to the energy efficiency of the property can be reflected in an increase in rent levels. This economic incentive, however, must not be at the expense of the tenant. As guidance, the total cost of rent and heating should stay the same. Thus any rent increase would be offset by the reduction in energy consumption (ignoring the value of the additional warmth). It may be necessary to protect tenants from exorbitant rent increases, particularly in properties where the rent was already high for a relatively inefficient dwelling, and, in specified circumstances, to any landlords with grants to bridge the gap between the cost and returns.

Both of these concerns imply a level of regulation in privately and, perhaps, publicly rented accommodation. The Group is reluctant to propose a new administrative burden, but is concerned about the implications both for investment and for rents in an unregulated market.

References

Boardman, B. (1993) Energy Efficiency incentives and UK households, *Energy and Environment*, 4, No. 4, Multi-Science, pp. 316–34.

Boardman, B. (1992) Social aspects of energy efficiency, in Christie, I. and Ritchie, N. (eds), *Energy Efficiency*, Policy Studies Institute, London, pp. 71–95.

Energy Saving Trust (1994), *Recommendations on the Standards of Performance in Energy Efficiency for the Regional Electricity Companies*, report for the Office of Electricity Regulation.

Further reading

Barker, T. and Johnstone, N. (1993) Equity and efficiency in policies to reduce carbon emissions in the domestic sector, *Energy and the Environment*, Multi-Science, Brentwood, Essex.

Boardman, B. (1990) *Fuel Poverty and the Greenhouse Effect*, Friends of the Earth, Heatwise Glasgow, Neighbourhood Energy Action, National Right to Fuel Campaign.

Boardman, B. (1991) *Ten Years Cold*, Neighbourhood Energy Action, Newcastle.

Boardman, B. and Houghton, T. (1991) *Poverty and Power: the efficient use of electricity in low-income households*, Bristol Energy Centre, Bristol.

Brown, A. (ed) (1992) *The UK Environment*, Department of the Environment, HMSO.

Burridge, R. and Ormandy, D. (1993) *Unhealthy Housing: Research, remedies and reform*, E & FN Spon, London.

Cm 2427 (1994) *Climate Change: the UK programme*, HMSO.

DoE (1993), *Energy Efficiency in Council Housing: Interim guidance for local authorities*, Department of the Environment.

DUKES (1993) *Digest of UK Energy Statistics*, HMSO.

HC648-I (1993) *Energy Efficiency in Buildings*, Environment Committee, House of Commons, Session 1992–3, Vol I, Report, HMSO.

HC648-II (1993) *Energy Efficiency in Buildings*, Environment Committee, House of Commons, Session 1992-3, Vol II, Minutes of evidence, HMSO.

HC648-III (1993) *Energy Efficiency in Buildings*, Environment Committee, House of Commons, Session 1992-3, Vol III, Minutes of evidence, HMSO.

Moore, R. (1991) *Cold, damp and mouldy housing in England*, paper presented at Unhealthy Housing Conference, Warwick.

NACAB (1993) *Out of Control?*, National Association of Citizens Advice Bureaux, London.

Oseland, N.A. and Ward, D.D. (1993) *An interim evaluation of the Home Energy Efficiency Scheme (HEES)*, limited circulation publication produced by BRE, but cited in Hansard and deposited in the House of Commons Library.

Pearson, M.A. and Smith, S.R. (1990) *Taxation and environmental policy: some initial evidence*, IFS Commentary no 19, London, Institute for Fiscal Studies.

Shorrock, I.D. and Bown, J.H.F. (1993) *Domestic Energy Fact File 1993 Update*, Building Research Establishment, Watford.

Ward, D. (1992), *Housing management and maintenance savings through energy efficiency refurbishment*, BRECSU, Building Research Establishment, Watford PD29/92.

Woodhouse, Peter R., Khaw, Kay-Tee and Plummer, Martyn (1993) Seasonal variation of blood pressure and its relationship to ambient temperature in an elderly population, *Journal of Hypertension*, 11, 1267–74.

Yarrow, G. (1992), *British electricity prices since privatization*, Regulatory Policy Institute, Studies in regulation 1, Hertford College, University of Oxford.

Yarrow, G. (1993), *Regulation and pricing performance in the gas industry*, Regulatory Policy Institute, Studies in regulation 2, Hertford College, University of Oxford.

Administrative issues and mechanisms 7

Ann Marno

Ann Marno

7.1 Introduction

This chapter looks at the issues and mechanisms which would be involved in the administration of a programme of work to upgrade the thermal and heating qualities of housing stock in Britain. It deals with existing housing, specifically; however, the Group recognizes that there will also be instances where it may be more cost effective to provide new housing to tenants rather than to upgrade existing dwellings to NHER≥8.

Reference is made to the *Energy Conservation Bill 1993 – How it could Work* document, which has been built upon in the writing of this Chapter. The Group strongly supported the Energy Conservation Bill 1994, and was greatly disappointed that it fell at its last hearing in the Commons. The main difference in proposals between this Chapter and the Bill is that the Group recommends the use of information for every dwelling, rather than using an audit on a percentage of the housing stock. The reason for this is to keep information address-specific.

The process of energy auditing is described and the introduction of mandatory energy auditing is justified, after which the programme of work is set out for local authorities, followed by a monitoring programme for central government and further justification of measures adopted within the local authority programme. Although reference is made to the National Home Energy Rating Scheme and its software, it should be noted that this is only one of the three government recognized schemes in operation, all of which use the Building Research Establishment's Domestic Energy Model (BREDEM) in order to deliver the government's Standard Assessment Procedure (SAP).

It should also be noted that this Chapter is rather prescriptive in detailing a course of actions for implementing the Group's proposals. The Group had difficulty in agreeing a single methodology, and fully recognizes that the system which is used in this Chapter is only one of many possible ways of achieving the proposals. The Group was also aware, however, of the need to steer a course through the various issues involved.

Paramount among the issues of carrying out improvement works is the question of who should foot the bill. The cost for the improvement of the housing stock will have to come largely from a combination of central and local government, with a heavy weighting towards central government. This would appear to be quite appropriate, considering the fact that large proportions of the savings which would be achieved through adopting the Group's proposals would be in the sphere of health, which is a central government expenditure.

The next issue for consideration is the need to target those households on low incomes before others – thus attacking the root of the problem of fuel poverty. This requires audits to be carried out at some level on all dwellings in order to make them address-specific.

It is important to recognize that any programme of improvements cannot take place overnight. It is apparent, therefore, that such a programme should begin as soon as possible.

It is also important to recognize that improvements to public properties cannot occur successfully if they take place in a vacuum of public ignorance. Tenant and other community organizations should be fully consulted and be able to play a fully participative role in consideration of improvement works, as suggested by the Department of the Environment's publication *Energy Efficiency in Council Housing – Interim Guidance for Local Authorities* (revised 1993). The Group recognizes a number of very pragmatic reasons for consultation:

1. Tenants living in properties which are to be upgraded must be given information concerning what is happening since it will be happening to their homes. Furthermore, information on heating systems and controls and other advice will be required to be given to (and understood by) those living in the properties.
2. There may well be choices to make concerning fuel types, heating system types, heating controls etc. which are directly applicable to tenants, and which are best decided by tenants, as they will be living with those decisions afterwards.
3. It will be essential to involve tenants in any decision-making process regarding any phased improvements. The tenants will need to agree to the phasing, and will need to be confident that future improvement work will be carried out to time. It will be the tenants' own homes that will be under discussion, and tenant acceptance of and enthusiasm for improvement measures will naturally have a high correlation to their involvement in the specification and their understanding of those measures.
4. There is also a requirement to address the need for some form of compensation or heating credit for those households who will be in need of upgrading measures, but not in receipt of them for some time.

In this area, the Group acknowledges the relevance of the action proposed by the Campaign for Cold Weather Credits (CCWC); particularly given the remit of identifying the energy inefficient housing and low-income households. It would be unjust to identify people living in these combined situations, and not to offer help in the interim to those who have to wait for housing improvement.

The Group therefore suggests a similar system to that proposed by the CCWC – except operating by identified NHER rather than by geographical position (since this is incorporated into the NHER). Persons in receipt of HEES 'passport' benefits would be entitled to winter fuel credits depending on the identified NHER of their dwelling: e.g. £8 per week for NHER 0–1, £6 per week for NHER 1–2, etc. The amount of compensation paid is thus tied to the energy performance and location of the dwelling, once again requiring address-specific auditing. The Level 0 rating used to rate every dwelling is not accurate enough to give a specific rating per individual dwelling, but using the method adopted in this Chapter – that of labelling individual homes with the average energy rating of all the homes of their house type – will produce the required accuracy.

This will serve both to address any animosity which might otherwise arise between households which have and have not received improvements, and immediately to cure the worst symptoms of fuel poverty, with the causes to be eradicated later.

7.2 Energy auditing

It is imperative that the ability of dwellings to produce and store heat at an affordable cost to their occupants is measured in some manner (Chapter 8 discusses energy efficiency and cost). The best method of doing so is to energy audit and rate the stock, taking into account such details as climate, built form, insulation, heating systems, appliances and heating controls. Energy rating a dwelling is effectively marking its ability to carry out the function of keeping its inhabitants warm. For the purpose of this Chapter we make use of the NHER scheme which rates a dwelling from 0 (very poor) to 10 (excellent).

It is proposed that all dwellings of NHER ≤ 2 be upgraded to a value ≥ 8. Those dwellings with the lowest ratings are to be targeted first because they are in the greatest need of improvement. It may be possible to increase the ratings of all properties incrementally; but for a real solution to fuel poverty and a reduction in global warming, complete programmes of improvements must be carried out as soon as is practically possible.

The Group accepts the fact that many local authorities will desire to help as many householders as possible by a smaller amount, rather than 'purging' a small number of homes (although the intention of the programme is to purge a large number of homes). If this is the case, then it may be possible to achieve the full NHER improvement in a sequence of two or more stages.

It will be important to note, however, that phased improvements are likely to be more expensive than a full package 'installed' at one time. Phased improvements would be likely to produce more disturbance to the householder, and can also produce other problems, e.g. a home with a rating of 0–1 will have little or no insulation, an inefficient heating system and an expensive fuel. The most cost-effective way of raising such a home's rating may well appear to be to introduce an efficient heating system using cheap fuel. However, that home would require a far smaller and far cheaper system if it were fully insulated. If the insulation were to be carried out at a later date, the decision makers would be faced with a quandary. The heating system could be of a size that it adequately heats the dwelling as it is, and is then far too big when insulation is added to the property at a later date; or the size might suit the dwelling as it would be with insulation, in which case it would be far too small to give correct comfort temperatures for the uninsulated house as it stands.

In the first instance, a far greater outlay would be made for a heating system which would probably be less efficient to operate than its cheaper alternative. It will also be the case that tenants cannot afford to use full central heating all day. In the second case, the tenant would have to suffer lower temperatures, or use some form of temporary (and probably expensive) heating until such time as the insulation is installed.

It is proposed, therefore, that if some form of phasing is chosen, then where possible the phased measures should equate to the package as a whole. For example, if it is found to be cheapest (and still effective) to heat a dwelling with two gas wall heaters (as it may well be with effective insulation) then these should be installed rather than a full central heating system. Such a measure is unlikely to be popular with tenants, however, unless they are fully consulted and given a proper explanation of the phasing programme. They will then appreciate that the first step is only in the short term and that the next phase (i.e. insulation) will result in a substantial improvement.

The Group would recommend that, wherever applicable, fuel credit should still be paid for properties which are only partially upgraded, and that extra rent should not be charged until NHER has been raised sufficiently to ensure that, even taking into account tenant use of comfort gains, the tenants will be saving at least enough to pay for the rent increases.

It will be important to undertake some form of mass energy audit of all residential buildings – of whatever tenure – in order to identify accurately the worst properties and the most applicable and cost effective improvements. The prime reason behind this mass audit is for a common ground to be established, so that all housing all over the nation can be compared, like with like, taking into account all the factors that affect energy usage. Without such a mass audit this comparison cannot be made and any distribution of funding to improve the situation would be subject to disparities caused by lack of information, nor would it stand up to an objective view of the most useful targeting of resources.

The audit will need to cover homes which are owned by local authorities (LHa) those owned by other, semi-public organisations such as housing associations, (HAs); those owned by private landlords (PLs) and those which are owner-occupied (OO).

One organization, however, must be responsible for carrying out all the audits in each area, to ensure a line of responsibility and that the process continues swiftly and effectively. The logical choice for such an organization is the local authority. Local authorities have been responsible for the construction of a large fraction of the nation's housing. They continue to be responsible for housing control, and are therefore required to carry a degree of information on all sector properties within their area. Thus they are the most likely source of information on all existing housing within their area. Local authorities will have to use whatever means are at their disposal, e.g. estate agents' records etc., in order to gather information on privately owned dwellings.

In the first instance, an NHER Level 0 rating would be carried out in order to identify broadly those properties most in need of attention. Local authorities will undertake the Level 0 survey, which should largely be possible using housing stock records, either computerized or filed.

The items requiring identification for **Level 0** analysis are:

- Built form
- Age
- Number of storeys
- Number of rooms
- Wall insulation
- Loft insulation
- Floor insulation
- Glazing (frame and type)
- Heating system type and fuel
- Water heating type and fuel.

(In the case of flats, the exposure of the roof and the floor must also be stipulated.)

It is also recommended that the following data items be added to the Level 0 list. Some of the data is geographical, to enable correct comparison of housing stocks from different areas (Level 0 information was developed for use within a single HA or LA, not for comparing one LA with another) and the remaining data would make costing of improvement work far more accurate:

- Degree-day region
- Height above sea level
- Outer wall construction
- Dwelling floor area
- Secondary heating type and fuel (if readily available).

Combining the extra information with the Level 0 analysis would produce a level of information which, for the purposes of this Chapter, we shall call **Level 0 plus**.

The only items which would not necessarily be in LA records would be the secondary heating system and fuel type. If this information is not readily available, then the audit may proceed without it, or assumptions could be made.

Gaps in the knowledge of LAs about their own stock can be filled using informed assumptions made by the authorities' housing management who know the stock best. Other information may come from other areas – for example Home Energy Efficiency Scheme (HEES) providers could be cross-checked by the authority for the addresses of properties in which loft insulation has been improved and draughtproofing carried out (although the Group recognizes that this practice may incur difficulties regarding confidentiality).

Gaps pertaining to residential properties of other tenures can be filled directly from the parties involved. Just as LAs will be obliged to provide central government with information, so should HAs be obliged to provide the information on their stock to LAs on a 3–6 monthly basis, with a great deal of information exchanged initially, followed by a lesser amount concerning improvements and new properties thereafter.

It may not be possible to place private landlords under any duty of information to the LA. The alternative could be to carry out a percentage audit during each Local House Condition Survey performed by the LA (as suggested by *Energy Conservation Bill – How it could Work*), and also to treat the tenants of private landlords in a similar way to owner-occupiers, in terms of data gathering.

Thus tenants of private landlords, together with owner-occupiers, would be sent a questionnaire to fill out for each address. Together with the form would be information on how and where to find the basic information necessary for the Level 0 plus audit, with some advice on what to do in problem cases. It will also be important at this stage to provide a telephone backup service to ensure that forms are correctly completed. This telephone information could be readily available via the Local Energy Advice Centres (LEACs), or via telephone advice lines run separately by each LA, or groups of smaller LAs, or on a central advice lines supported by central government.

The start of the information campaign could coincide, perhaps, with media coverage explaining why the audits are being carried out, and how they will serve to reduce fuel poverty and to help the environment, as a means of ensuring public support and speed in returning forms.

By these means, address-specific information for the entire housing stock can be gathered and used. Computer systems which are capable of dealing with this amount of information are now readily available; and address-specific information will be needed for future use – both for deciding which dwellings to upgrade first, and for deciding what extra income for fuel any home should have thereafter.

7.3 The need to make energy auditing mandatory

Many local authorities have already carried out energy audits on their own housing stock. In many cases, the primary reason that these audits were carried out was to secure funding which would otherwise have been forfeit, but now the importance of the information made available by the auditing process has proven useful.

Unfortunately, many authorities have not yet carried out audits – often because they have been unable to afford the cost of software and training which their larger or perhaps more visionary neighbours have undertaken. This disparity is likely to continue if energy auditing is not made obligatory for all. For a mass audit to be complete, it must be mandatory rather than voluntary. It may also be the case that LAs cannot be laid under a duty of law to undertake energy auditing of private stock. If this should prove to be the case, then they could be laid under a 'duty of funding' to do so, as they effectively were for their own stock under the Green House Programme.

The issue of whether or not LAs can afford to pay for the rating of the housing stock within their area has not been addressed here. Most LAs with large volumes of their own properties have already done so; however, this process has not included rating private dwellings, which is likely to incur greater difficulty. Furthermore, LAs with smaller housing stocks have often not used energy auditing, due to lack of resources.

Considering the economic benefits of carrying out the auditing work in each area, it is unlikely that funding will be necessary for this activity. It may be necessary to offer some form of 'up front' funding in cases where the LA has only a small housing stock and is operating in an area which includes a large number of private dwellings.

The fact that some local authorities are 'ahead of the field' in terms of energy auditing may serve to make the introduction of mass energy improvement work somewhat easier. This is a point which has not been addressed by this chapter, but it should be borne in mind that an enormous quantity of training and gearing up will be necessary in the relevant industries in order to support such a major round of improvement works.

7.4 Programme of work

Energy auditing is only part of the process of classifying and upgrading properties. It is necessary in order to identify those properties which are most in need of improvement, and in order to identify the most cost-effective measures for improving those properties.

The Level 0 plus rating used here will not identify the NHER of any dwelling on its own. It will give a statistically correct rating for a number of dwellings, but will not give a fully accurate rating for individual dwellings. The first purpose of the Level 0 plus audit is to identify the worst areas of stock in each local authority. The second purpose is to identify the numbers involved to central government for the purpose of funding to each local authority.

The seven main stages of the programme are as follows.

1. **Level 0 plus information will be gathered on all residential housing stock.** That will be accomplished through local authorities using their own

expertise or that of consultants, based on information gathered from:

- Local authority records and management knowledge for public sector rented accommodation.
- Housing associations for their own housing stock. HAs will be obliged to provide this information on a 3–6 monthly basis.
- Self-administered questionnaires completed by occupants of privately rented and privately-owned stock (and any other means available).

2. **Local authorities will prepare databases of information on the four types of residential housing within their area: LA, HA, PL, OO.** The databases should be capable of retaining information from a full audit for every address, though at this stage they may contain little more information than that required for Level 0 plus. The local authority will then energy audit the entire housing stock within their area to Level 0 plus. The databases and the results of the Level 0 plus audit will be sent to central government on a 3–6 monthly basis, as and when they become available.

The LAs will then identify all housing types and, using Level 0 plus analysis, discover the average NHER of each. Every dwelling of a particular house type will then be labelled on the database with an NHER equal to the integer below that of the house type's average value. This will have the purpose of identifying households on one of the qualifying benefits. There are two reasons for doing this: first, to focus on those and second because any extra income for fuel would be paid through Housing Benefit as this is the most all-encompassing benefit, and is already address-specific. The address is important, since as soon as recipients move, their extra fuel income also change; Housing Benefit would also change when any move occurred, whereas other benefits would not.

An important point to make is that as any extra fuel income would be paid in cash, the system will be open to 'abuse'. The abuse is possible both by recipients and by later governments. Recipients might well spend the additional money on commodities other than fuel. Later central governments, however, might abuse the system by allowing other areas of benefits to fall in real terms and relying upon recipients to change their spending from necessary fuel expenditure in order to supplement other needs.

It should also be recognized that the benefits to the nation (e.g. reduced health costs) of this scheme would not necessarily accrue from this programme of work if either form of abuse were to occur.

The only way in which such abuse may be guarded against would be to pay all forms of fuel income as address-specific fuel tokens. Recipients would not be able to spend the income on anything other than fuel; and it would be clear what element of fuel income was being given to those in receipt of benefits.

The Group recognizes, however, that this would be a very paternalistic approach, and one which has been argued against by many organizations, including the CCWC. It can be argued, for example, that all forms of cash benefit are open to such abuse, that it may not be the best thing for recipients (some of whom undoubtedly already suffer colder conditions in order to subsidize other necessities) and that it is against freedom of choice.

In order to prevent abuse of the system by central government, therefore, the Group suggests that the government makes clear what amount of any benefit is regarded as being allocated to fuel costs.

Abuse of the system by recipients can best be countered by improving the state of the housing stock with all possible speed; since after such improvement, the extra fuel benefit would no longer be payable.

3. **The LAs should then identify the worst housing types within their housing stock.** These should be the house types which consistently give NHERs of ≤2. When the LA has chosen properties for upgrading work a Level 0 plus audit of the selected properties should have an average rating of ≤2, with not more than 10% of the properties indicated to be above 2.

After central government has confirmed acceptance of the house types identified for improvement, LAs should begin their process of tenant consultation. During and as a part of such consultation, the LAs should select a representative sample of each house type, of varying size, and perform detailed audits on these properties. The purpose of the detailed audits will be to cost a number of packages of measures which would be acceptable to tenants, to improve the NHERs to ≥8 (if a number of packages should prove to be possible), assuming economies of scale etc. Each audit should take 40–60 minutes on average. Analysis will take far longer, to ensure that cost-effective measures are looked at in detail, including, for example, checks for condensation.

It will then be possible for the costs of the packages to be correlated with respect to house size etc. The LA will then produce a plan of upgrading work for all the identified LA and HA stock in order to improve it to an NHER ≥8. In some cases it may be possible to perform the upgrading work in stages, in order to give some immediate benefit to more householders.

Upgrading to ≥8, will ensure that the household will still gain from having an improvement, despite losing its winter fuel credits and paying a higher rent.

4. **The results of costed analyses will be sent to central government for confirmation of analysis and monitoring purposes.**

Costs for certain areas will be higher than for others. This will be due to dispersal of the stock and weather conditions. Work undertaken in Scottish Island communities, for example, will suffer high costs for transporting materials and labour, as will most rural areas in comparison with towns and cities. Furthermore, large variations can occur in the cost of wall insulation. Cavity wall insulation is an excellent and very cheap way of insulating properties which have cavity walls. It would not be applicable, however, in an area with a very low incidence of cavity walls, or in many western areas which suffer very heavy rainfall – particularly Wales, South-West England and Western Scotland. The need for wall insulation in these areas, however, may well exceed the need in others, and so will necessitate the use of external insulation and cladding. This is a cost which may seem high under traditional analysis but which would be cost effective if a wider cost-benefit analysis were used (Chapter 8); and which must be borne if real benefits are to occur.

5. **Central government funding will be given for improvement work to public housing**. Some funding from LAs and certainly from HAs may be required but this could be subject to negotiation at a later date. It will almost certainly be the case that some areas will receive more central government funding than others. This may seem unfair, but it is the only way for all the lowest rated dwellings to be attended to first. For this reason, and for others already given, funding should be mostly derived from central government, which is concerned with the welfare of all citizens equally, irrespective of location.

Tenants and homeowners in the PL and OO stock with NHER ≤ 2 will be offered the chance to make use of long-term central government loans at 0% interest. The loans will be address-specific: they will be tied to the address, not to the tenants/homeowners at that time. This will relieve tenants and homeowners of the anxiety that any work carried out to the property would not pay for itself until long after they have vacated the property.

Any tenants/homeowners offered such a loan would have to perform an energy audit (the cost of which would be incorporated into the loan) showing what can be done to the house and that it would be cost-effective – using a pay-back of less than 15 years. Loans would be repayable over a period of up to 15 years, depending upon the size of the loan and the yearly savings. The maximum annual repayment would be that of the estimated fuel cost savings.

6. **Work will commence on upgrading properties**. In public sector housing, priority will go to those households which are known to be on low incomes and spending long periods inside their homes, i.e. those who would benefit most from the improvements and who would reduce the outlay of central government by the greatest amount.

As public properties are upgraded, tenants in receipt of housing benefit will lose their extra fuel income which they formerly received. However, this loss will be more than matched by the drop in necessary fuel expenditure. Furthermore, the rents for all public upgraded properties will rise. A suggested value for extra rent would be half of the equivalent fuel cost savings in moving from NHER 5 to 8. It will provide the LAs and HAs with funds to recoup their share of the expenditure on the housing, and will go some way to providing investment for further improvements elsewhere in the stock.

Similarly, as work is completed on those PL and OO properties which have been given loans for improvement, so would any households formerly in receipt of extra fuel benefit lose it.

As work progresses, all information databases will have to be kept up to date. It would be obligatory for information from the detailed audits of privately owned properties in receipt of central government loans to be supplied to the relevant LA, which would then report back to central government every 3–6 months on the improvement work of all sectors.

7. **The programme of work will begin again for those properties with ratings between 2 and 4.** This will enable established working practices and working teams to remain without any break in continuity, which will achieve maximum return from the setting-up costs, training of staff, etc. After the cycle for properties ≤4, this will be followed by a programme to improve remaining dwellings with NHER ≤6. It is expected that the first two programmes will take longest, as they probably contain greater numbers of dwellings. A suitable time-scale would be six years for each of the first two programmes, followed by a final programme lasting approximately four years.

This process may appear restrictive but the Group recognizes that the worst housing needs to be addressed first. It may be the case that, for example, this process will result in estates being revisited; but unless this is accepted, then the very worst housing will not be addressed within an acceptable time-scale.

7.5 Job creation/ training

The benefits in terms of retraining and job creation would be considerable. Most LAs would require a number of surveyors, and also permanent staff to use the computing side of the auditing system. The main benefits in terms of job creation, however, will be in the building sector – which is greatly in need of such a boost – and in the manufacture of insulation and efficient heating systems and appliances.

The jobs so created will not be short-term: in fact, they are likely to continue in the main for the next 10–16 years and so may be considered ideal for retraining/re-skilling schemes for many of the older unemployed who require some form of work or career for this length of time.

7.6 Monitoring by central government

Considering the scale of investment which the programme of work involves, an adequate monitoring system will be imperative. Central government will need to monitor four main areas:

1. That data collection for energy auditing has been carried out fully and to stipulated quality standards.
2. That cost estimates used by LAs are reasonable and accurately describe the availability of resources and staff within their areas.
3. That work is carried out to a quality standard.
4. That databases are maintained, over time, to the stipulated quality standards.

These monitoring areas should be addressed in the following ways:

1. At present there is no quality standard for Level 0 information provision. Such a standard is being devised by the National Energy Foundation in conjunction with its members and Energy Advisory Services Ltd., who

wrote the NHER suite of programmes. Such a standard would be necessary. However, Level 0 plus information should be provided in an upgradable form (i.e. in the NHER scheme, not the stand-alone NHER Level 0 and Stockprofiler programmes, but one such as Auto-evaluator, which gives Level 0 assessment but can be readily upgraded) and in the same format by each local authority.

Furthermore, central government should publish guidelines developed to deal with certain factors arising from Level 0 plus information acquisition, similar to the conventions used within the NHER itself (e.g. if floor insulation has not been specifically noted in the records of a property, then that property should be deemed to have none, as it is very unlikely that it would have any).

In the case of the NHER, this is presently being addressed. Such a system of guidelines must be established in order to ensure that the computer records for all housing stock information sent to central government will be format-checked. Any not corresponding to the correct format will be returned and funding withheld until the correct formats are presented.

At the more detailed energy audit level, qualified assessors would be needed to carry out the work. These assessors should be accredited in some way to give energy audits, e.g. the National Energy Foundation carries out its own quality control system on NHER Assessors.

2. Validated photocopies of estimates etc. should be provided along with all other information on costings. These would then be checked if found to be incorrect in an accounting sense, or seemingly too expensive or too cheap.

3. The HEES scheme provides a quality standard for loft insulation, draught-proofing and hot-water tank insulation. Cavity wall insulation standards are provided by the Board of Agrément. Similar standards would be necessary for internal dry-lining insulation and external cladding insulation. British Standards would be required for internal and external insulation. The quality of heating system installation would be covered by CORGI membership of all installers of gas systems; and membership of a similar electrical contractor's organization would be required of installers of electrical heating.

4. Similar standards would be used to those for (1.). Lessons can be learned from the Energy Action Grants Agency (EAGA) from the administration of HEES. LA 3–6-monthly energy reports would be cross-checked with those for households receiving extra fuel benefit, and with the addresses of dwellings in receipt of central government loans for upgrading work.

7.7
Conclusion   The programme of work proposed here builds upon the Energy Conservation Bill 1994 and the programme of action petitioned for by the Campaign for Cold Weather Credits, seeing both as excellent schemes and two sides of the

same coin.

The programme of auditing will use statistically accurate, broad-brush auditing – rating the housing stock in order to identify order of improvement. Funding requirements for LAs would, in the main come from central government.

Auditing will be address-specific, for use with a system of increased fuel credit payments linked to housing benefit for those households in greatest need until such time as they have improvements.

After identification, a sample of the worst public housing types will be audited in detail for the purpose of costing packages of improvements and ensuring their practicality. Local authorities will produce programmes of action for improving public dwellings from NHER ≤2 to NHER ≥8.

Owner-occupiers and private sector tenants will be eligible for 0% interest loans which will be tied to the property and repayable over a period of up to 15 years.

As improvements are completed on dwellings, their residents will no longer be eligible for extra fuel credits and will either pay slightly increased rents, or repay loans from central government.

References

The Energy Conservation Bill 1993 – How it could Work, Association for the Conservation of Energy.

Energy Efficiency in Council Housing – Interim Guidance for Local Authorities, revised 1993, The Department of the Environment.

Cost implications of a policy for affordable warmth

8

Bill Sheldrick

Warmth is a fundamental prerequisite for the well-being of society. However, as seen in the evidence collected by the Watt Committee, and well documented elsewhere, a significant proportion of our dwelling stock is neither warm, nor dry, nor affordable to heat. In Scotland, the 1991 *Scottish House Condition Survey* found one in four dwellings to be suffering from dampness, condensation or mould growth problems (Scottish Homes, 1993).

Within the domestic sector, the fundamental aim of our energy policy should be to ensure that the basic needs and standards of society are being met. If minimum standards are not being met, then policies and programmes should be designed to redress this situation. If energy policy, when combined with the heating and insulation characteristics of the dwelling stock, cannot ensure that homes are warm, dry and affordable to heat, then there is something wrong with the policy. If despite the evidence the situation is not rectified, then something is wrong with priorities. Rather than redress this situation, aspects of the present energy policy (e.g. VAT on fuel) will exacerbate the existing situation.

Given the present heating and insulation characteristics of the dwelling stock, achieving adequate levels of warmth amongst many low-income households will require an increase in energy consumption at a time when the government's emphasis is on reducing energy consumption to achieve a reduction in carbon dioxide emissions. However, warmth and energy efficiency are not mutually exclusive objectives, and to present them as such is a false dichotomy. Taking effective action to realize an affordable warmth policy, and to reduce the level of energy consumption, can be pursued in tandem, through investing in the heating and insulation characteristics of the dwelling stock (Boardman, 1991). To achieve these dual objectives will not be cheap, but then the cost paid by those people living in cold, damp, mouldy and expensive to heat housing is already very high (see Hunt and Boardman, Chapter 2).

The debate about energy efficiency in the home tends to be dominated by the concerns of an economic paradigm: too often the rhetoric is that insulation and heating programmes have to be justified in terms of the cost effectiveness of the investment, the internal rate of return, or the pay-back periods. This would be acceptable if a truly 'economic' assessment was performed and the impact on the utility function (i.e. the general welfare of society) was evaluated. For example, Treasury guidance on assessing the returns on investment states (DoE, 1978):

> where the landlord is a public body all the benefits should be counted, including those to the tenant, because what matters in the expenditure of public funds is simply whether the total sum of benefits is greater than that of costs, whoever receives them.

The problem of investment appraisal is not with the theory but with the practice: economic theory gets replaced by financial accounting where the prime concern is money. Thus the cost to a household living in a cold, damp and mouldy home is paid for through poor living conditions and deprivation, and with health. Yet when the benefits of an insulation investment programme are calculated, too often only the value of the reduced energy consumption is included in the equation with other benefits ignored or even derided. If affordable warmth is to be a policy objective, these wider matters and concerns have to be incorporated into the equation. Continuing to ignore them will only result in further misery for a significant proportion of the population and hidden costs to the nation.

8.2
The cost of warmth versus energy consumption

Fuel bills and house temperatures are the result of the interaction between a variety of climatic, construction, heating and behavioural factors, and fuel prices. Any household is not better or worse off, financially or comfortwise, simply because it uses more or less fuel than another. This can be illustrated through the results of an energy audit of an inter-war, three-bedroom, semi-detached house in Glasgow, with minimal levels of insulation. Five different heating systems were assessed with regard to meeting a range of **demand temperatures** (that is, the temperature households are striving to achieve through the use of heating rather than the actual or average temperature achieved within the dwelling):

- Gas fires.
- A wall-mounted gas boiler with radiator system (not fan assisted).
- A condensing gas boiler with radiator system.
- Direct-acting electric heaters charged on the peak rate domestic tariff.
- An electric storage radiator system charged on an off-peak tariff.

The energy audits demonstrate a different result depending on whether it is in the cost of heating or the energy consumption being examined (Figure 8.1). In terms of energy consumption (i.e. gigajoules), the direct-acting

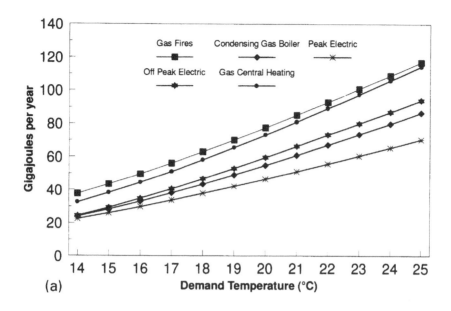

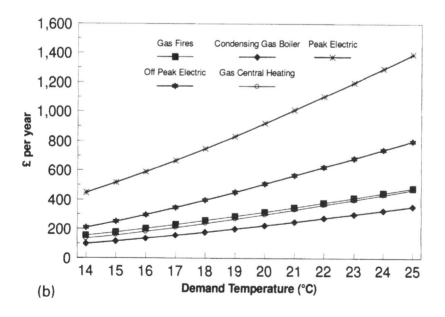

Figure 8.1 (a) Energy consumption for heating (GJ); (b) cost of energy consumption for heating (£).

electric heaters are the most efficient of those systems assessed in meeting a demand temperature. However, they are also considerably more expensive to use than any of the other heating systems assessed here. The condensing gas boiler system and the off-peak storage system consume about the same amount of energy in meeting the dwelling demand temperature, but the gas condensing boiler does so at a much lower cost to the household. Heating by gas fires, which consumed the most energy in meeting the demand temperatures, was the second cheapest in terms of meeting that demand.

This difference between energy consumption and the cost of warmth is a function of the price of fuel and the efficiency of the heating appliance at converting fuel into heat. The most efficient heating systems are not necessarily the cheapest to run. Direct-acting electric heating charged on the standard domestic tariff is almost five times more expensive than gas, and over twice as expensive as off-peak electricity in terms of the unit price of fuel. The difference between gas and electricity is reduced when the concept of a 'useful unit of heat' is used (i.e. taking into consideration the amount of fuel that a heating appliance or system needs to use to deliver that heat) (Table 8.1). In terms of a 'useful unit of heat', direct-acting electric heating is about four times more expensive than a gas condensing boiler system, and about 2.5 times more expensive than a gas fire; off-peak electric heating is only 20% more than a gas fire and 30% more expensive than the more common gas boiler types. For low-income households, the cost of a 'useful unit of heat' reduces the confusion arising from the variability in fuel prices and appliance efficiencies by focusing on the issue of concern, i.e. warmth.

Table 8.1 Cost of Fuel – unit prices and cost per useful unit of heat

Heating appliance	Unit price[1] (pence per kWh)	Efficiency[2]	Cost of a useful unit of heat (pence per kWh)
off-peak storage radiator (day)	7.70p	100%	7.70p
direct-acting electric heater	7.13p	100%	7.13p
off-peak storage radiator (night)	2.93p	100%	2.93p
pre-1979 floor-mounted gas boiler	1.48p	55%	2.69p
gas fire	1.48p	60%	2.47p
wall-mounted gas boiler	1.48p	65%	2.28p
gas combined boiler	1.48p	65%	2.28p
fan assisted gas boiler	1.48p	68%	2.18p
condensing gas boiler	1.48p	85%	1.74p

1 prices in effect in Glasgow in March 1994
2 system efficiencies utilized in the NHER Homerater programme (version 2.6)

As with energy consumption, one cannot determine whether a household is purchasing sufficient warmth simply from the size of its fuel bill. A low fuel bill may indicate a dwelling that is heated inadequately. Alternatively, it could reflect a very well insulated dwelling with a highly efficient boiler using a low-cost fuel. The cost of heating a dwelling can vary quite considerably (even when occupant behaviour is excluded) depending on the heating system installed and the level of insulation.

8.3 Cost of warmth

Using the same three-bedroom, semi-detached dwelling as in Section 8.2, the cost of heating to different demand temperatures was calculated using the NHER Homerater programme (version 2.6) assuming different levels of insulation and variations in the installed heating appliances (set out below). The results are presented in Figure 8.2 and compare the impact of single improvements and the packages of heating and insulation improvements respectively on the Base Case dwelling in terms of the cost of heating:

BOX 8.1   SINGLE IMPROVEMENTS

1. Base case: the unimproved dwelling with a gas fire in the lounge and reliance upon direct-acting electric heating appliances elsewhere in the dwelling. The actual dwelling did not have any installed heating other than the gas fire in the lounge. The dwelling has minimal amounts of insulation: 50 mm of loft insulation and the external doors have been draughtproofed, but not the windows.
2. Double glazing (D/G): assumes that wooden framed, double glazed units with integral draughtproofing have been installed.
3. Cavity wall insulation (CWI): assumes that the dwelling has had its cavity walls insulated with mineral fibre.
4. Gas fires only: assumes that the whole house is heated by gas fires.
5. Gas central heating: assumes that a wall-mounted gas fired boiler, with radiators and a room thermostat, programmer and thermostatic radiator valves, has been installed.
6. Electric off-peak heating (Off/Pk): assumes that an electric storage radiator system with automatic charge control on an off-peak tariff has been installed to provide 90% of the heat into the lounge (with the other 10% from a direct acting electric heater) and 100% elsewhere in the house.

PACKAGES OF IMPROVEMENTS

1. Base case: as above.
2. Basic insulation improvement (HEES): assumes the house has received basic insulation measures funded under the Home Energy Efficiency Scheme, i.e. draughtproofing of all windows and external doors, 150 mm of loft insulation, and an 80 mm hot-water tank jacket.

3. Basic insulation plus heating (HEES + Htg): assumes the installation of basic insulation measures as set out above, as well as gas fires throughout the house, in keeping with the affordable warmth scheme proposed as one of initial three 'E-factor' pilot programmes (Ofgas, 1992).

4. Insulation plus heating (HEES,CWI + Htg): assumes that cavity wall insulation is included along with the basic insulation and heating package as set out above.

5. Building regulation standards (BRegs + Htg): assumes that thermal standards of the dwelling complied with those set out in the 1990 and 1991 Building Regulations for new dwellings and that the heating was provided by the gas central heating system set out above.

6. Super insulation and condensing boiler (Super + Condensing): assumes that loft has a U-value of 0.2, the walls 0.35, and the floor 0.25; that double glazing has been installed and that the heating and hot water are provided by a condensing gas boiler.

7. Super insulation and electric storage radiators (Super + Off/Pk): assumes the same insulation standards as above but that the heating is supplied by Scottish Power's 'Comfort Plus' off-peak electric heating and hot water system.

The cost of heating to any demand temperature is dependent on (amongst other factors) the levels of insulation, the form of heating and the cost of fuel. As can be seen in Figure 8.3 the cost of heating can range quite significantly. As the insulation standards increase and the efficency of the heating system improves, the cost of heating the dwelling to a given temperature reduces. For a demand temperature of 21 °C, the cost of heating ranged from £795 per year in the Base Case, to £330 per year amongst the single improvement measures, and down to £70 per year with the packages of improvements assessed.

Spending £330 per year on heating (Line X in Figure 8.3a) would purchase a demand temperature of less than 14 °C in the Base Case situation (Line Y in Figure 8.3a), but would be sufficient to heat the dwelling throughout to a demand temperature of 21 °C if there was a gas central heating system installed (Figure 8.3a). The impact of the packages of improvements on the cost of heating is such that it is as cheap or cheaper to heat the dwelling to a demand temperature of 25 °C where the insulation standards meet or better those of the 1990/91 Building Regulations, regardless of whether a gas central heating or an electric storage heating system is installed, as it is to heat the existing dwelling to a demand temperature of 14 °C given the use of a gas fire in the lounge and direct-acting electric heaters elsewhere in the dwelling (Line A in Figure 8.3b).

The difference in the cost of heating this dwelling to a demand temperature

Cost per year (£)

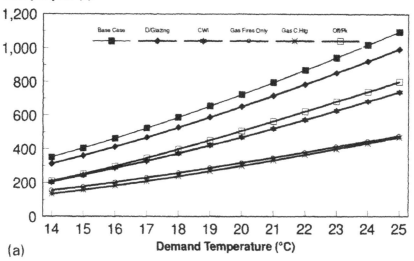

(a)

Cost per year (£)

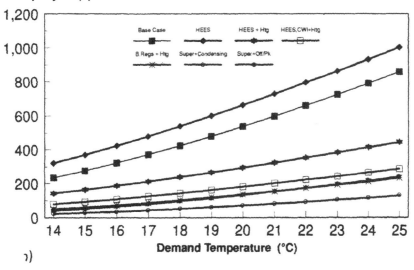

ı)

8.2 Annual heating costs. (a) Single improvements; (b) improvement
ıs. (See Box 8.1 for definitions.)

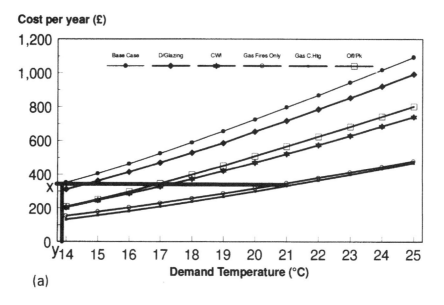

(a)

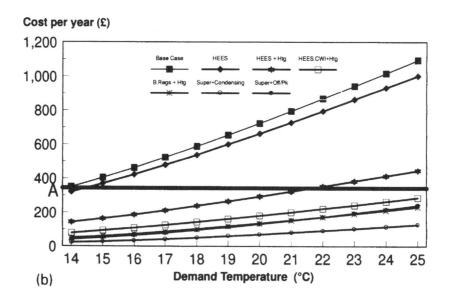

(b)

Figure 8.3 Cost of warmth. (a) Single improvements; (b) improvement packages. (See Box 8.1 for definitions.)

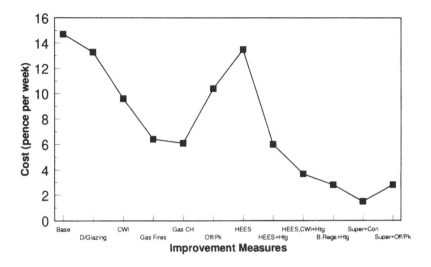

Figure 8.4 Cost of heating (per m² of total floor area per week). (See Box 8.1 for definitions.)

for the situations examined can be further demonstrated by examining the average weekly cost of heating per m² of floor area. (The dwelling comprises approximately 104 m² of floor area in total.) For the Base Case, the estimated annual heating costs represents an expenditure 14.7 pence per m² of floor area per week (Figure 8.4). This cost falls to 1.3 pence per week with the super insulation and the condensing boiler package. The other improvement measures, both single measures and packages of improvements, fall somewhere in between these extremes for this dwelling.

**8.4
Cost of
adequate
warmth**

Hunt and Boardman (Chapter 2) outline a heating regime of a demand temperature of 21 °C in the lounge and 18 °C in the rest of the house for 13 hours per day. The Energy Efficiency Office (1985) has noted 'that these temperatures are commonly found in well-heated homes.' For many low-income households, this standard will represent a significant improvement on those found within their homes. Temperature monitoring in dwellings with poor insulation coupled with expensive-to-use forms of heating has shown that, despite spending significantly more than average on fuel, households can still be medically at risk from cold homes (Sheldrick, 1994). While this temperature standard will represent an increase for many low income households it may not be sufficient for all, for as the Energy Efficiency Office also noted, 'The elderly, sick, disabled, or very young children may need slightly higher temperatures' (Energy Efficiency Office, 1985).

A distinction is drawn here betweeen the demand temperature, the actual temperature and the average internal temperature, as they are not the same. The **demand temperature**, as set out above, is that temperature that households are trying to achieve when the heating is in use. When the heating is on, the **actual temperature** is dependent on the external temperature and the capacity of the appliances, e.g. a gas or electric heating system that complies with the British Standard is designed to meet 21 °C in the lounge and 18 °C elsewhere in the house when the external air temperature is −1 °C. The **actual internal temperature** may be less than the demand temperature when the dwelling is warming up or the external temperature is extremely cold. Alternatively, the demand temperature may be exceeded as a result of internal gains or overheating with a badly controlled heating system.

The **mean internal temperature** differs from the demand temperature in that it is concerned with the periods when the heating is off as well as those when it is on. As a result, the mean internal temperature will reflect the insulation characteristics of the dwelling; the poorer the insulation, the faster and further the internal temperature will drop for a given external air temperature. Figure 8.5 sets out the resultant mean internal temperatures profiles for the lounge and the rest of the house respectively for varying levels of insulation (assuming the existing heating appliances). As the insulation specification increases, the resultant mean internal temperatures also rise. For many low-income households, this situation will not be one of overheating their homes, but one where they are moving closer towards temperature standards that are considered reasonable.

In addition to identifying a heating pattern, Hunt and Boardman also set a minimum acceptable mean internal house temperature standard of 14.5°C. This minimum standard is consistent with the National Home Energy Rating (NHER) scheme convention of declaring a dwelling to be inadequately heated when the annual mean internal temperatures achieved with its fixed heating is calculated to be less than 14.5 °C (National Energy Foundation, 1993). Figure 8.6 identifies the demand temperature at which the mean internal temperatures within the dwelling, given the different levels of insulation, fall below 14.5 °C.

It can be seen in Figure 8.6 that the poorer the insulation within the dwelling , the higher the demand temperature that is needed to ensure that the dwelling maintains an adequate mean internal temperature: the heat that is put into the dwelling is not retained when the heating is not used and therefore the internal temperature drops faster and further than in an insulated dwelling. As the insulation levels improve, the demand temperature required for the dwelling to be adequately heated drops.

The Base Case dwelling requires a demand temperature of 20 °C to achieve adequate temperatures across the whole house. The installation of basic insulation and double glazing increases the mean internal temperatures within the house, but is still insufficient in itself to breach the 14.5 °C threshold for any demand temperatures below 20 °C. Other insulation measures have a

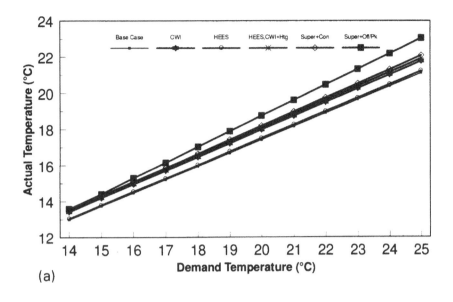

(a)

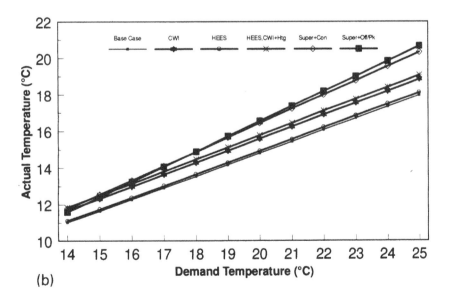

(b)

Figure 8.5 Resultant mean internal temperatures (a) Lounge; (b) rest of the house. (See Box 8.1 for definitions.)

Cost per year (£)

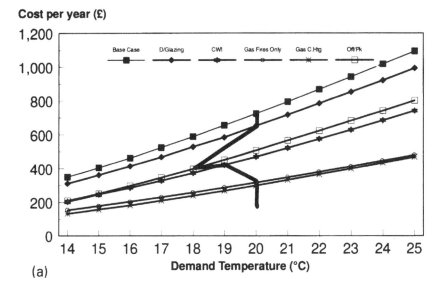

(a)

Cost per year (£)

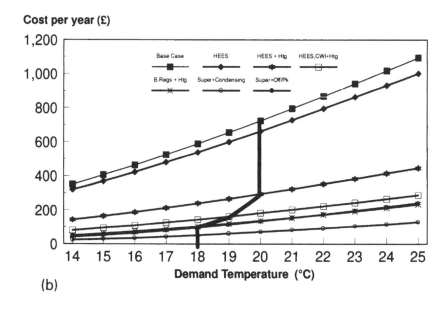

(b)

Figure 8.6 Adequate heating costs. (a) Single improvements; (b) improvement packages. (See Box 8.1 for definitions.)

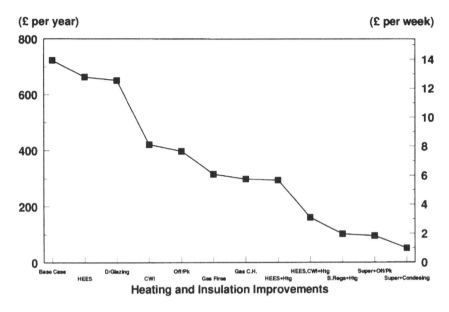

Figure 8.7 Cost of adequate heating. (See Box 8.1 for definitions.)

larger impact: the required demand temperature falls to 19 °C with cavity wall insulation (Figure 8.6a), and to 18 °C with insulation to the 1990/91 Building Regulation thermal standards and the super insulation package (Figure 8.6b). The effect of reducing the demand temperature is to reduce the cost associated with heating the dwelling to an adequate temperature by reducing the amount of fuel required to be consumed. A similar impact can be achieved with improving the heating system; if not by reducing the demand temperature, then by allowing the same demand temperature to be realized for a lower cost.

The annual cost of heating the dwelling to an adequate temperature standard as calculated for the various heating and insulation packages is set out in Figure 8.7. For the Base Case, the level of expenditure required simply to achieve an adequate temperature standard is in almost £14 per week; the same dwelling, but built to the thermal standards of the 1990/91 Building Regulations, would require an expenditure of about £2 per week on heating to achieve adequate temperatures across the whole house.

8.5 Energy rating

The indicative energy ratings of the Base Case dwelling and all the various heating and insulation scenarios were calculated using the NHER scheme. The NHER is not concerned solely with energy consumption, but with that aspect of fuel consumption that most focuses a household's attention – the fuel bill. The NHER is an energy cost index on a scale of 0 to 10, derived from the total annual fuel costs per m^2 of total floor area (using standard assumptions about heating patterns and occupant behaviour). Thus the NHER score not only takes account of space heating costs, but also those for water heating, cooking, lights and appliance use, and standing charges. The use of standard assumptions and relating NHER score to a m^2 of floor area allows different sized dwellings and different household sizes to be compared directly.

The Base Case dwelling scores 2.9 on the NHER scale. While this NHER rating is about 2 points below the national average, it is higher than would be achieved for the same dwelling if it was without any loft insulation (i.e. an NHER of 2.5), without loft insulation and a hot-water cylinder jacket (i.e. an NHER of 2.2). If the dwelling was reliant completely on direct-acting electric heating, it would score 0.7 on the NHER scale. This NHER score would fall further to 0.3 if the dwelling was without any insulation.

The NHER for this dwelling does increase with the various heating and insulation improvements, rising to an NHER score of 8.9 with the super insulation package and a condensing gas boiler (Figure 8.8a). Except for the installation of gas central heating, all of the single measures assessed and basic HEES insulation packages increase the NHER score by 1 point or less on the NHER scale. The impact varies between the different improvements assessed. The biggest increases occur with the comprehensive insulation and heating packages and with significant changes in the heating system (e.g. installing super insulation plus condensing boiler increases the NHER by 6.2; meeting the Building Regulation standards plus gas central heating, 4.8; gas central heating, 3.2; installing comprehensive insulation plus gas fires throughout, 3.2).

In Figure 8.8b the costs of the various single measures and the improvement packages are plotted against the incremental increase in the NHER. (The costs of the various improvements are outlined in section 8.7 below). It is evident that small increases in the NHER can be had for comparatively small investments in heating and insulation, but that large jumps require improvements to both the heating and insulation characteristics of the dwelling.

**8.6
A standard
for affordable
warmth**

The previous sections have examined separately four approaches to categorizing domestic energy use: energy consumption, cost of warmth, adequate heating, and energy rating of a dwelling. The suitability of each with regards to defining affordable warmth varies.

Of the four, energy consumption *per se* is the least useful. This can be best illustrated with two graphs. In Figure 8.9a, the energy consumption per m^2 of the dwelling is plotted for the different situations assessed. Some improvements increase energy consumption, some reduce it. Insulation improvements on their own will reduce energy consumption, but where they are coupled with changes in the heating, energy consumption can either increase or decrease.

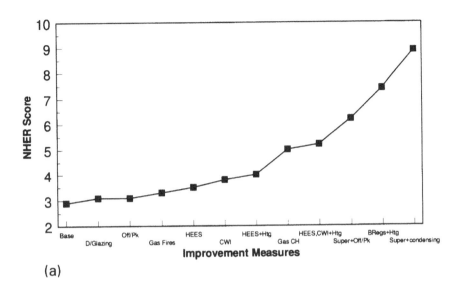

(a)

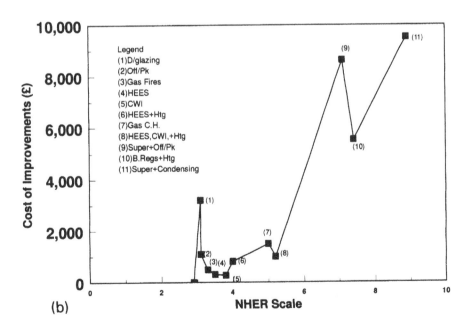

(b)

Figure 8.8   (a) Increase in NHER score; (b) cost of increase in NHER score.
(See Box 8.1 for definitions.)

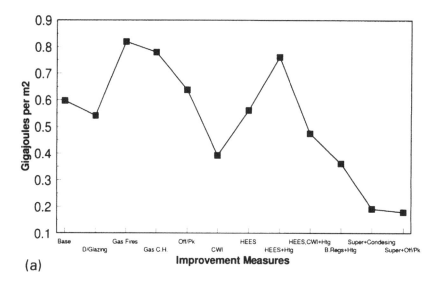

(a)

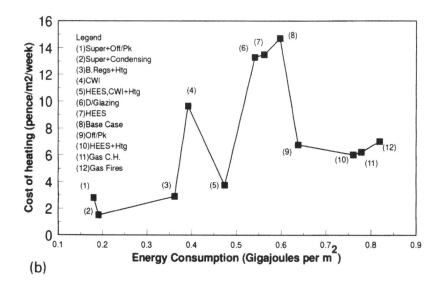

(b)

Figure 8.9   (a) Energy consumption (GJ/m^2); (b) cost of energy consumption (pence per GJ per m^2 per week). (See Box 8.1 for definitions.)

In Figure 8.9b, energy consumption per m² is plotted against the cost of heating per m² per week. All of the improvements reduce the cost of heating compared with the Base Case. However, switching from direct-acting electric heating to gas fires throughout the dwelling (with or without insulation) or switching to gas central heating will result in an increase in energy consumption. The other improvements assessed reduce not only the cost of heating but also the energy consumption.

While energy consumption can be ruled out as an affordable warmth indicator, the cost of heating, adequate heating and the NHER could all have a rôle to play in defining an affordable warmth standard, either individually or as part of a composite measure. Criteria for defining affordable warmth could include:

• total annual/weekly fuel cost per household/dwelling/m²;
• total annual/weekly heating cost per household/dwelling/m²;
• minimum temperature standard;
• a minimum NHER score;
• percentage of income on heating/total fuel.

If dwellings are going to be upgraded so that they are affordable to heat to a reasonable and acceptable temperature standard across all income groups, then it is necessary to set a high standard for heating and insulation improvements. The Group has decided to advocate an NHER score of 8 as the standard to be aimed for when dwellings are upgraded. To achieve such a target will require action to be taken on upgrading both the insulation and heating characteristics of the dwelling.

8.7 Cost of achieving a standard

The standard of affordable warmth adopted will determine the extent of the improvements required and the cost. If nothing is done, then the investment cost will be nil (the cost implications will be experienced in other expenditure budgets such as health). Every other strategy has a cost implication, the amount being dependent on the extent of the improvements. Continuing only with a basic insulation improvement programme may not cost much per dwelling, but then the potential benefits are only minimal in terms of both heating cost reductions and improved temperatures. Despite being targeted at those on low incomes, the HEES scheme is not about achieving affordable warmth.

The basis of the costs used for calculating the level of investment required for the various heating and insulation improvements is as follows:

• Basic insulation (i.e. draughtproofing, loft insulation): £315 per dwelling (based on HEES grant).
• Basic insulation plus heating (gas fires throughout): £815 per dwelling (Energy Saving Trust proposals).
• Basic insulation, cavity wall insulation and gas fires: £1000 per dwelling (Energy Savings Trust proposals).

- Cavity wall insulation: 94 m² @ £3 per m² = £280 (Energy Saving Trust proposals).
- Double glazing: 16 m² @ £200 per m² = £3200 (from NHER Cost File).
- External cladding: 94 m² @ £35 per m² = £3290 (from Heatwise Glasgow quotes).
- Gas central heating: £1500 (from Heatwise Glasgow quotes).
- Condensing boiler: £2000 (from Heatwise Glasgow quotes).
- Floor insulation: 51.8 m² @ £8.50 per m² = £440 (from NHER Cost File).
- Gas fires: £500 per dwelling (from Energy Savings Trust proposals).

When grants are available or proposed, the amount of grant available has been used as the cost. Experience has shown that prices will gravitate (even downwards) towards the allowable grant, particularly when the grants are well monitored.

Using the above figures, the cost of the different single measures and packages of improvements assessed in the energy audits are set out in Table 8.2.

In Figure 8.10a, the reductions in the cost of heating the dwelling to a demand temperature of 21 °C are plotted for each of the heating and insulation measures assessed. While all of the measures reduce the cost of heating, their financial pay-back varies: those measures with the largest reductions are not necessarily those with the best pay-back periods (Figure 8.10b). Those measures with good performance indicators are also not necessarily those with the best market profile or consumer preference – cavity wall insulation has a much greater impact on reducing the cost of heating and a much lower pay-back period than double glazing, but has a much lower rate of uptake than does double glazing (Sheldrick, 1993).

Further, large investments do not necessarily reap large reductions in heating costs (Figure 8:11). Double glazing and basic insulation improvements have similar (small) impacts on fuel bills but have very different capital costs.

Table 8.2   Cost of heating and insulation improvements

double glazing	£3260
cavity wall insulation	£280
gas fires	£500
gas central heating	£1500
off peak electric heating	£1100
basic insulation	£315
basic insulation plus heating	£815
basic insulation, cavity wall insulation and gas fires	£1000
building regulation standard plus gas central heating	£5545
super insulation plus condensing boiler	£9525
super insulation plus off peak heating	£8625

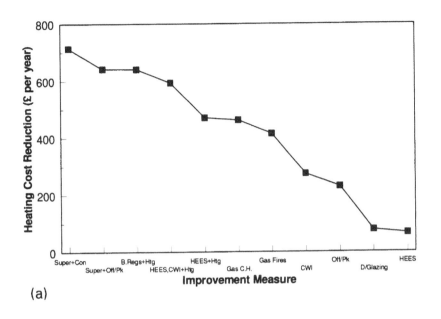

(a)

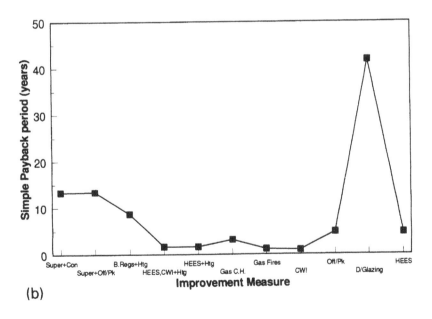

(b)

Figure 8.10   (a) Reduction in heating costs; (b) payback period on heating cost reduction. (See Box 8.1 for definitions.)

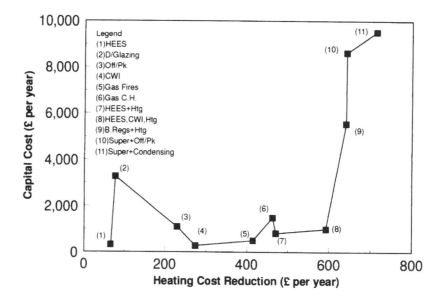

Figure 8.11 Heating cost reduction vs capital cost. (See Box 8.1 for definitions.)

Double glazing is the most costly single measure on the list of improvements assessed, and would never be cost effective on its own in energy terms, but that does not stop householders installing it and paying large sums of money to do so.

When the various measures are assessed in terms of the capital cost for a one-point increase on the NHER scale, the majority of the improvements have a similar cost. On this assessment, installing gas fires is less effective than installing a super insulation and gas condensing boiler package despite a difference in capital costs of over £8500 (Figure 8.12). The highest cost per increase in the NHER score is associated with installing double glazing, and then gas fires, while the lowest costs are for cavity wall insulation and the HEES improvements respectively.

As stated at the start of this section, the cost of investment will depend upon the standard of affordable warmth to be achieved. The largest reductions in fuel bills within the option assessed here have high investment costs. To achieve an NHER score of 8 will not be cheap. For the Base Case dwelling, an NHER score of 8 could be achieved through a package of improvements comprising gas central heating system, 150 mm of loft insulation, 50 mm of floor insulation, cavity wall insulation and double glazing, for a cost of around £6,000 in Glasgow. To achieve the same NHER target in the more populated areas of England would cost about half as much, because the inclusion of double glazing within the specification is not needed.

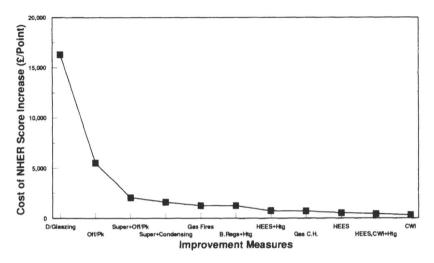

Figure 8.12   Cost of increase in NHER score. (See Box 8.1 for definitions.)

Low-cost improvements alone or only those measures with quick pay-back periods will not necessarily achieve the affordable warmth target for low income households. It was because 'economic' criteria such as pay-back periods and cost effectiveness were not delivering affordable warmth in terms of average fuel costs or as a percentage of disposable income that Glasgow City Council adopted its Action for Warm Homes strategy (Glasgow City Council, 1990 and 1993). This strategy has as its goal the largest reduction in household fuel bills for the investment funding available. To achieve this strategy has required a re-ordering of investment priorities. Not all of the investment may be considered cost effective under traditional investment criteria, but when all of the impacts are considered Glasgow City Council believes that the strategy represents a very high value for money invested.

**8.8
Heating
additions**

While improving the insulation and heating standards within a dwelling is one approach to realizing lower heating costs for low-income households, another is the use of fuel subsidies. In the past, heating additions have been paid out for difficult to heat dwellings, for estates with disproportionately expensive to use heating systems, and for central heating (amongst others). The only subsidy at present is £7 per week under the Severe Weather Payment Scheme. Even if this subsidy was paid out every week of the year (i.e. £312 at 1993 rates), it would not reduce the cost of heating in the Base Case house in Figure 8.1 by as much as insulating the house to the 1990/91 Building Regulation standards and installing gas central heating. Severe Weather Payments are not, however intended as a general subsidy, but as an exceptional needs payments. For the payments to be triggered, the average external air temperature has to fall to 0 °C or below for seven consecutive days (at the designated monitoring station). Those eligible in Scotland for Severe Weather Payments, which is not all low-income households, are likely to receive such payments more regularly than claimants in England because of the generally colder Scottish climate. This subsidy, although useful, can only have a limited impact, as in most years there are only one or two qualifying weeks.

By defining a target standard for expenditure on heating, the Annual Heating Cost graphs (Figure 8.2) can be used to determine the level of heating subsidy that would be needed to achieve both adequate and reasonable temperatures for each dwelling. To illustrate this approach, a target heating expenditure of £385 per year is set, representing a mean weekly expenditure of £7 per week (derived from 60% of the average household expenditure on fuel in Scotland). Figure 8.13 shows that households would qualify for a heating subsidy when the cost of achieving the demand temperature necessary to keep the mean internal temperature above 14.5 °C (Figure 8.4) fell within the shaded area of the graph. Expenditure below this level, and therefore a lower demand temperature, would be the householder's responsibility, while heating to a higher demand temperature, and the additional cost incurred, would be the householder's own choice.

For the Base Case, the heating subsidy would be £410 per year. Where the dwelling had gas central heating, no subsidy would be payable. Figure 8.13 shows that the dwelling would qualify for a heating subsidy even if it had double glazing, or gas fires throughout, or had received basic insulation improvements under HEES. These measures on their own are not sufficient to realize adequate warmth for an average expenditure on fuel.

If the target temperatures are to be achieved a system of credits such as that devised by the Cold Weather Credits Campaign (1992) might be used.

While the heating subsidy of £410 calculated above on the balance to achieve a reasonable demand temperature is less than the cost of installing cavity wall insulation to the same dwelling (i.e. £280), installing cavity wall insulation in the dwelling would reduce the annual heating bill and the level of heating subsidy. This one-off investment in cavity wall insulation would pay for itself through the saving in the subsidy in less than 2 years. The alternative is to pay a continuing subsidy or for the cost to be paid through health problems, deteriorating fabric and deprivation. Unfortunately welfare subsidies are the responsibility of the Department of Social Security, and investment in housing and energy efficiency is the responsibility of the Department of the Environment and The Scottish Office.

**8.10
Conclusion**

This examination of the cost implication of affordable warmth has used one particular dwelling as a sample. It is not an a typical dwelling with an unusual construction nor is it an extreme case. The same process can be carried out with various house types to calculate an affordable warmth investment

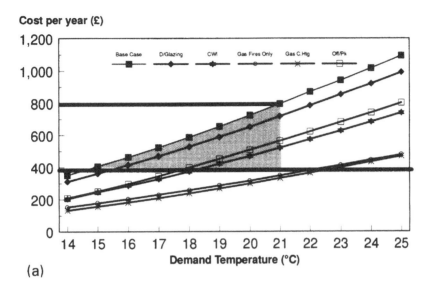

(a)

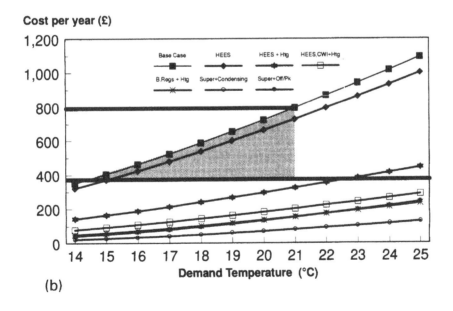

(b)

Figure 8.13 Heating addition costs. (a) Single improvements; (b) improvement packages (See Box 8.1 for definitions.)

profile, given the breakdown of the dwelling stock and its heating and insulation characteristics, to achieve an affordable warmth standard. With sufficiently detailed information on the breakdown of the UK housing stock and the levels of insulation, a similar analysis could be carried out and then aggregated to allow for a good estimate of the cost of achieving an affordable warmth policy, whether through investment or through subsidizing fuel costs. Some of this information was available to the Watt Committee, but not enough to allow for detailed estimates to be prepared. However, it was seen from the data available that the necessary amount runs to over £20 billion.

Acknowledgement   The basis of some of this work is in Cold, condensation and housing poverty by Thomas A. Markus in *Unhealthy Housing: Research, remedies and reform*, (eds Roger Burridge and David Ormondy), 1993, E & FN Spon, London.

References   Boardman, B. (1991) *Fuel Poverty*, Bellhaven Press, London.
Coldweather Credits Campaign (1992) *Warm Homes in Cold Weather*, Coldweather Credits Campaign, Glasgow.
Department of the Environment (1978) Remedial work for existing electrically heated dwellings. *Domestic Energy Note 3*, Working Party on Heating and Energy Conservation, Department of the Environment, London.
Energy Efficiency Office (1985) *Guide to Home Heating Costs*, Department of Energy, London.
Energy Efficiency Office (1993) Benefits to the landlord of energy efficient housing. *Good Practice Case Study no. 155*, Department of the Environment, London.
Glasgow City Council (1990) *Glasgow Action for Warm Homes*, City Housing, Glasgow City Council, Glasgow.
Glasgow City Council (1990) *Glasgow Action for Warm Homes: Critical path*, Ellis Fisher Award Submission, City Housing, Glasgow City Council, Glasgow.
National Energy Foundation (1993) *NHER Homerater Manual*, National Energy Foundation, Milton Keynes.
Ofgas (1992) *Gas and Energy Efficiency – The "E" Factor*, Ofgas, London.
Scottish Homes (1993) *Scottish House Condition Survey 1991: Survey Report*, Scottish Homes, Edinburgh.
Sheldrick, B. (1985) *Hard to Heat Estates: A review of policy and practice*. Working Paper 267/10.85, Social Policy Research Unit, University of York, York.
Sheldrick, B. (1993) *Energy efficiency and the domestic sector*, Task 4: Scottish Energy Study, A.H.S. Emstar and Scottish Enterprise, Glasgow.
Sheldrick, B. (1994) Energy and fuel consumption, in *Poor and Paying for It: The price of living on a low income*, (ed. G. Fyfe), Scottish Consumer Council, HMSO, Edinburgh.

Source material

Association for the Conservation of Energy (1994) *The Fifth Fuel: the newsletter of the ACE*, No. 28, Spring 1994, ACE, London. Brennan, P., Greenberg, G. Miall, W. and Thompson, S (1982) Seasonal variation in arterial blood pressure, *Brit Med J* ii 919–923.

British Gas, consumption data concerning gas-fired tumble driers, British Gas, Holborn, London, 1994.

Bull, G. and Morton, J. (1978) Environment, temperature and death rates. *Age and Ageing 7* 210–224.

Burr, M.L., Dean, B.V. Merrett, T.G. et al. (1980) Effect of anti-mite measures on children with mite sensitive asthma: a controlled trial. *Thorax* **35** 506–12.

Callaghan, J. (1978) Text of speech to the Coal Industry Society (14 December, 1977), in Energy conservation: a series of speeches by politicians and industry leaders, Department of Energy, London.

Collins, K.J. (1986) The health of the elderly in low indoor temperatures. Paper presented at the Conference on Unhealthy Housing: a diagnosis. University of Warwick, 14th–16th December.

Coull, B.M., Breamer, N., De Garmo, P., Sexton, G., North, F., Knox, R., et al (1991) Chronic blood hyperviscosity in subjects with acute stroke, transient ischaemic attack and risk factors for stroke. *Stroke* **22** 162–168.

Dales R.E., Zwanenburg H., Burnett R., and Franklin C.A. (1991a) Respiratory health effects of home dampness and molds among Canadian children. *American Journal of Epidemiology* **134** 196–203.

Dales, R.E., Burnett, R. and Zwanenburg, H. (1991b) Adverse health effects among adults exposed to home dampness and molds. *American Journal of Epidemiology* **134** 505–510.

Department of Environment (1992) *Climate Change*, DoE, London.

Douglas, A. and Rawles, J. (1989) Excess winter mortality: the cardiovascular component, Paper presented at Meeting of the Scottish Society of Experimental Medicine, 3rd November, Aberdeen.

Environment Select Committee (November 1993) *Report on Energy Efficiency in Buildings*, HC 648-1-111, HMSO.

Environment Select Committee (March 1994) *Energy Efficiency: the role of OFGAS* Minutes of Evidence, HC 328–i, HMSO.

Hansard (1993) WA col. 492 1st April.

Hopton, J. and Hunt, S.M. (1990) Changes in health as a consequence of changes in housing. Paper delivered at the Society for Social Medicine Conference, Glasgow.

Hosen, H. (1978) Moulds in allergy. *J Asthma Res* **15** 151–6.

Hunt, S.M., Martin, C.J. and Platt, S.P. (1988) *Damp Housing, Mould Growth and Health Status. Part I* Report to the funding bodies. Glasgow and Edinburgh District Councils.

Hunt, S.M. and Lewis (1988) *Damp Housing Mould Growth and Health Status Part II* Report to the funding bodies. ibid.

Hyndman, S.J. (1990) Housing dampness and health among British Bengalis in East London. *Soc Sci Med* 30 131–41.

Kannel, W., Wolf. P., Castelli, W., and D'Angostino, R. (1987) Fibrinogen and risk of cardiovascular disease: the Framingham study. *J Amer Med Ass* 258 1183–16. Keithley.

Keatinge, W., Coleshaw, S., Cotter, F. et al (1984) Increases in platelet and red cell counts, blood viscosity and arterial pressure during mild surface cooling: factors in mortality from coronary and cerebral thrombosis in winter. *Brit Med J* 1 1405–1408.

Lloyd, E. (1986) *Hypothermia and Cold Stress,* Croom Helm. London.

Markus, T.A. and Morris, E.N. (1980) *Buildings, Climate and Energy,* Pitman, London.

Meade, T.W., Mellows, S., Brozovic, M., Miller, G., Chakrabati, T., North, W.R. et al (1986) Haemostatic function and ischaemic heart disease: principal results of the Northwich Park heart study. *Lancet* ii 533.537.

Mintour, P. (1994) Minister talks out conservation bill, *The Guardian*. 23/4/94, page 4.

Northup, S. and Kilburn, K. (1978) The role of mycotoxins in human pulmonary disease. In *Mycotoxic Fungi and Mycotoxicosis. Vol 3: Mycotoxicosis of Man and Plants.* Academic Press. London.

Platt, S.P., Martin, C.J. and Hunt, S.M. (1989) Damp housing, mould growth and symptomatic health state. *BMJ*, 298, 1673–78.

Platt-Mills, T. and Chapman, M. (1987) Dust mites: immunology, allergic disease and environmental control. *Journal of Allergy & Clinical Immunology*, 80, 744-75.

Qizibash, N., Jones, L., Warlow, C. and Mann, J. (1991) Fibrinogen and lipid concentration as risk factor for transient ischaemic attacks and minor ischaemic strokes. *BMJ* 303 605–9.

Salonen, J.K., Puska, P., Kottke, T.E. et al. (1983) Decline in mortality from coronary heart disease in Finland from 1969–1979. *BMJ* i 1857–60.

Scarisbrick, C., Newborough, M. and Probert, D. Improving the thermal performances of domestic electric ovens, *Applied Energy*, 39, 263–300, 1991.

Scottish Energy Study (19xx).

Scottish Home and Health Department (1988) Scottish Health Service Costs. Information Services Division. Common Services Agency. Edinburgh.

Smith, J.E. and Moss, M.O. (1985) *Mycotoxins: Formation, Analysis and Significance,* John Wiley and Sons Ltd., Chichester.

Social Security Statistics (1993) HMSO, London.

Social Trends (1994) Central Statistical Office. HMSO, London.

Tunstall-Pedoe, H. (1989) Heart disease mortality. *Brit Med. J.* i 751–2.

Tuomilheto, J., Geboers, J., Salonen, J.T. et al. (1986) Decline in cardiovascular mortality in North Karelia and other parts of Finland, *BMJ* ii 1068–76.

Tobin, R., Baranowski, E. Gilman, A. et al. (1987) Significance of fungi in indoor air: report of a working party. *Can J Pub Health* 78 Suppl. 1–14.

Woodhouse, P., Khaw, Kay-Tee and Plummer, M. (1993) Seasonal variation of blood pressure and its relation to ambient temperature in an elderly population. *Journal of Hypertension* 11 1267–74.

World Health Organisation (1987) *Health Impact on Low Indoor Temperatures,* WHO, Copenhagen.

Appendix
The Watt Committee
on Energy

1. The objectives of the Watt Committee on Energy are:

   (a) to promote and assist research and development and other scientific or technological work concerning all aspects of energy;
   (b) to disseminate knowledge generally concerning energy;
   (c) to promote the formation of informed opinion on matters concerned with energy;
   (d) to encourge constructive analysis of questions concerning energy as an aid to strategic planning for the benefit of the public at large.

2. The concept of the Watt Committee as a channel for discussion of questions concerning energy in the professional institutions was suggested by Sir William Hawthorne in response to the energy price 'shocks' of 1973/74. The Watt Committee's first meeting was held in 1976; it became a company limited by guarantee in 1978 and a registered charity in 1980. The name 'Watt Committee' commemorates James Watt (1736–1819), the great pioneer of the steam engine and of the conversion of heat to power.

3. The members of the Watt Committee are 50 British professional institutions. It is run by an Executive on which all member institutions are represented on a rota basis. It is an independent voluntary body, and through its member institutions it represents half a million professionally qualified people in a wide range of disciplines.

4. The following are the main aims of the Watt Committee:

   (a) To make practical use of the skills and knowledge available in the member institutions for the improvement of the human condition by means of the rational use of energy.
   (b) To study the winning, conversion, transmission and utilization of energy, concentrating on the United Kingdom but recognizing overseas implications.
   (c) To contribute to the formulation of national energy policies.

(d) To identify particular topics for study and to appoint qualified persons to conduct such studies.

(e) To organize conferences and meetings for discussion of matters concerning energy as a means of encouraging discussion by the member institutions and the public at large.

(f) To publish reports on matters concerning energy.

(g) To state the considered view of the Watt Committee on aspects of energy from time to time for promulgation to the member institutions, central and local government, commerce, industry and the general public as contributions to public debate.

(h) To collaborate with member institutions and other bodies for the foregoing purposes both to avoid overlapping and to maximize co-operation.

5. Reports have been published on a number of topics of public interest. Notable among these are *Air Pollution*; *Acid Rain and the Environment*; *The Chernobyl Accident and its Implications for the United Kingdom*; *Nuclear Energy: a Professional Assessment*; *Technological Responses to the Greenhouse Effect*; *Rational Use of Energy and the Environmental Benefits*; and *Methane Emissions*. Others are in preparation.

6. Those who serve on the Executive, working groups and sub-committees or who contribute in any way to the Watt Committee's activities do so in their independent personal capacities without remuneration to assist with these objectives.

7. The Watt Committee's activities are coordinated by a small permanent secretariat. Its income is generated by its activities and supplemented by grants by public, charitable, industrial and commercial sponsors.

8. The latest Annual Report and a copy of the Memorandum and Articles of Association of the Watt Committee on Energy may be obtained on application to the Secretary.

Enquiries to:
The Information Manager,
The Watt Committee on Energy,
40 Grosvenor Place,
London SW1X 7AE.
071 233 2565

British Association for the
  Advancement of Science
British Nuclear Energy Society*
British Wind Energy Association
Chartered Institute of Building
Chartered Institute of Management
  Accountants*
Chartered Institute of Purchasing
  and Supply
Chartered Institute of Transport
Chartered Institution of Building
  Services Engineers
Combustion Institute (British
  Section)
Geological Society of London
Hotel Catering and Institutional
  Mangagement Association
Institute of Biology
Institute of British Foundrymen
Institute of Chartered Foresters
Institute of Energy*
Institute of Home Economics
Institute of Hospital Engineering
Institute of Internal Auditors (UK)
Institute of Management Services
Institute of Marine Engineers
Institute of Materials
Institute of Mathematics and its
  Applications
Institute of Petroleum*
Institute of Physics

Institute of Refrigeration
Institute of Wastes Management
Institution of Agricultural
  Engineers
Institution of Engineering
  Designers
Institution of Environmental
  Sciences
Institution of Gas Engineers*
Institution of Mining and
  Metallurgy
Institution of Mining Engineers
Institution of Nuclear Engineers
Institution of Plant Engineers*
Institution of Structural Engineers
Institution of Solar Energy Society
  – UK Section
Operational Research Society
Royal Geographical Society
Royal Institute of British
  Architects*
Royal Institution of Chartered
  Surveyors
Royal Institution of Great Britain
Royal Institution of Naval
  Architects
Royal Society of Chemistry*
Royal Society of Health
Royal Town Planning Institute
Society of Dyers and Colourists
Textile Institute

**Member
institutions
of the Watt
Committee
on Energy**

* Denotes permanent member of the Watt Committee Executive

Publications of the Watt Committee on Energy

WATT COMMITTEE REPORTS

30 *Domestic energy and affordable warmth* 1994, xii + 141 pp. Illus. ISBN 0 419 20090 8 £39.95

27 *The rational use of energy and the environmental benefits* (papers presented at a conference in Strasbourg, 1-3 April 1992) 1992. Illus. 2 vol. £30.00 (per set)

26 *Profiting from low-grade heat* 1994 iv + 190 pp. Illus. ISBN 0 85296 835 5. £40.00

25 *Rational use of energy in urban regeneration* 1992. 200 pp. Illus. (Special issue of <u>Applied Energy</u> 43, 1–3) ISBN 0 946392 20 X £65.00

24 *Risk and the energy industries* 1992. iv + 130 pp. Illus. (Special issue of <u>Applied Energy</u> 42, 1&2) ISBN 0 946392 19 6 £80.00

23 *Technological responses to the greenhouse effect* 1990. x + 98 pp. Illus. ISBN 0 185166 543 9 £49.00

22 *Renewable energy sources* 1990. viii + 172 pp. Illus. ISBN 0 185166 500 5 £54.00

21 *The membrane alternative: energy implictions for industry* 1990. viii + 172 pp. Illus. ISBN 0 185166 476 9 £60.00

20 *Gasification: its role in future technological and economic development of the UK* 1989. x + 118 pp. Illus. ISBN 0 185166 326 6 £51.00

19 *The Chernobyl accident and its implications for the UK* 1988. xii + 148 pp. Illus. ISBN 0 185166 219 7 £49.00

18 *Air pollution, acid rain and the environment* 1988. x + 126pp. Illus. ISBN 0 185166 222 7 £49.00

17 *Passive solar energy in buildings* 1989. x + 70 pp. Illus.

ISBN 0 185166 280 4 £41.00

15 *Small-scale hydro power* 1985. vi + 62 pp. Illus. ISBN 0 946392 16 1 £28.00

14 *Acid rain* 1984. v+ 58 pp. Illus. ISBN 0 946392 15 3 £28.00

13 *Nuclear energy: a professional assessment* 1984. vi + 86 pp. Illus. ISBN 0 946392 14 5 £37.00

11 *The European energy scene* 1982. vi + 54 pp. Illus. ISBN 0 946392 12 9 £28.00

10 *Factors determining energy costs and an introduction to the influence of electronics* 1981. iv + 78 pp. Illus. ISBN 0 946392 11 0 £37.00

9 *Assessment of energy resources* 1981. iv + 84 pp. Illus. ISBN 0 946392 10 2 £37.00

7 *Towards an energy policy for transport* 1980. iv + 94 pp. Illus. ISBN 0 946392 08 0 £37.00

6 *Evaluation of energy use* 1979. iv + 112 pp. Illus. ISBN 0 946392 07 2 £37.00

5 *Energy from the biomass* 1979. iv + 76 pp. Illus. ISBN 0 946392 06 4 £28.00

4 *Energy development and land in the United Kingdom* 1979. vi + 50 pp. Illus. 6 full-colour maps. ISBN 0 946392 03 X (incl. 10 pp. supp. *Land for energy development*) £42.00

3 *The rational use of energy* 1978. iv + 72 pp. Illus. ISBN 0 946392 02 1 £28.00

2 *Deployment of national resources in the provision of energy in the UK, 1975–2025* 1977. 60 pp. Illus. ISBN 0 946392 01 3 £28.00

In Preparation: 28 *Methane emissions* (1993. Illus.); 29 *Combined heat and power* (1993. Illus.); Report numbers 1, 8, 12 and 16 are no longer available.

OTHER PUBLICATIONS

Five years after Chernobyl: 1986–1991
– A review (1991. x + 56 pp. Illus.).
ISBN 0946392 18 8 £20.00
Radiation monitoring around CEGB
nuclear power stations 1989. viii + 96
pp. Illus.
ISBN 0 946392 17 X £15.00
A Review of SERC sponsored
research work within the tertiary
education sector on energy
technology in the United Kingdom
(1989. iv + 66 pp.) £10.00

Index

Page numbers appearing in **bold** refer to figures and page numbers appearing in *italic* refer to tables.